# PRECALCULUS
# MATHEMATICS
# IN A NUTSHELL

# PRECALCULUS MATHEMATICS IN A NUTSHELL:

## GEOMETRY, ALGEBRA, TRIGONOMETRY

**George F. Simmons**
**Professor of Mathematics**
**Colorado College**

Janson Publications, Inc.   Providence, Rhode Island

Library of Congress Cataloging-in-Publication Data

Simmons, George Finlay, 1925-
  Precalculus mathematics in a nutshell.

  1. Mathematics—1961-    . I. Title.
QA39.2.S52   1987          512'.1          87-35300
ISBN 0-939765-13-6 (pbk.)

10  9  8  7  6  5  4

Printed in the United States of America

Designed and illustrated by John Johnson,
Teapot Graphics.

# PREFACE

For a variety of reasons, many bright people find it difficult to learn in high school the mathematics they need to know in order to continue effectively with their education. My purpose in this little book is to try to help by pulling together the essentials of the subject into three convenient small packages: geometry, algebra, and trigonometry. Most college mathematics teachers would be overjoyed (and astounded) if their students came to them from high school in full command of this much background material, but most high school graduates today do not even come close.

The three chapters that make up this book were written at different times with different aims. They are almost completely independent, and each can be mastered without reference to the other two. Chapter 3 contains several repetitions of earlier material which I have left in place so that students who wish to study only trigonometry can easily do so. Each chapter has many exercises, and answers are provided along with a number of fully worked-out solutions.

Precalculus mathematics is not an infinite shapeless mass that no one can hope to master. In order to demonstrate this, I have eliminated most of the repetitious and unnecessary material often included in precalculus mathematics textbooks. I have also tried to achieve the utmost brevity that will still permit students who are willing to work at it to understand the material. If I overstep the bounds here and there, and provide explanations that are too concentrated for clarity, I welcome suggestions from teachers and students for improvements that can be included in future editions.

Even though few students enjoy studying mathematics, many find it necessary to do so. If this book can occasionally ease the pain and smooth the learning process for these students, it will have done its job.

# CONTENTS

## CHAPTER 1. GEOMETRY

## CHAPTER 2. ALGEBRA

# CHAPTER 3. TRIGONOMETRY

# CHAPTER 1
# GEOMETRY

"Any fool can know. The point is to understand."
—Albert Einstein

# INTRODUCTION

Geometry is a very beautiful subject whose qualities of elegance, order and certainty have exerted a powerful attraction on the human mind for many centuries. The discoveries of Democritus and Archimedes about the volumes of cones and spheres (Sections 4 and 5 in this chapter) are among the most wonderful achievements of classical civilization. Also, the basic facts of geometry are absolutely essential for understanding many of the pure and applied sciences.

In spite of all this, most high school students emerge from their geometry courses with mixed feelings of confusion and relief. Why?

One of the reasons is that they have been ground down by complicated trivialities and offered little compensating insight into the geometric ideas that really matter. They have been bombarded with innumerable nit-picking definitions, and also with elaborate, boring "step-reason, step-reason" proofs of statements that in most cases are obvious to begin with. (At that stage, who can doubt the truth of such a statement as this: "Given any three points on a line, one is between the other two"? When asked to examine a proof, the natural reaction of an intelligent student is irritation and impatience, and he is right.) Their textbooks often seem to be written by the kind of person — we all know such people — who talks so much and says so little that we soon stop listening. All this tends to kill their interest in geometry long before they reach the meat of the subject.

The root of the problem is slavish adherence to the doctrine of Deductive Reasoning. This is the notion that knowledge is somehow not legitimate or genuine until it has been organized into an elaborate formal system of theorems that are carefully deduced from a small number of axioms or 'self-evident truths' stated at the beginning. Deductive Reasoning is an interesting idea that educated people ought to know something about, just as they should know something about representative government, the internal combustion engine, and other human inventions. It was very popular among philosophers and scientists of the 17th century, and was applied by them to physics,

2

ethics, and other unlikely subjects. Science shook off its grip 200 years ago, but geometry has continued to be strangled by this outmoded philosophical doctrine down to the present day.

In this chapter geometry is considered for its own sake and for the sake of its use as an indispensable tool in science and engineering, and *not* as a vehicle for teaching deductive reasoning. For us the purpose of proof is to remove doubt and convey insight, not to belabor the obvious. This point of view produces a gain in efficiency so great that almost everything of importance in plane and solid geometry can be said in about a dozen pages, with full explanations and proofs.* I have tried to include all the necessary facts and to omit everything else, however interesting or tempting it might be. (Sections 1 to 5 conform to this standard, but in Appendices E and F, I yielded to temptation and added a few delicious items just for the fun of it.) It should be added that no definitions are provided for such familiar geometric objects as triangles, parallel lines, circles, cones, spheres, and the like. This chapter is not intended to teach geometry to someone who knows absolutely nothing about it, but rather to review and clarify the main ideas of the subject; and if definitions are needed, they can be found in almost any dictionary.

I offer this bit of advice to the student. My explanations are deliberately very concise, with few words wasted. Also, much of the burden of the exposition is carried by the figures. Passive reading therefore will not do. If you wish to understand, it is necessary to read actively and carefully, thinking all the time, constantly asking *why?*—and constantly struggling to find an answer.

---

*I use the word "proof" to mean an argument that I hope my intended audience will find convincing. A few mathematicians may object to this relativistic attitude. However, since students differ from logicians in their power of skepticism, and logicians differ among themselves from one generation to the next, it seems unlikely that any fixed, unalterable, absolute meaning can possibly be attached to the concept of proof. What a proof is depends on who and when you are.

# 1. TRIANGLES

## (a) SUM OF ANGLES.

The customary unit of measure for angles is the degree. One degree (1°) is one-ninetieth of a right angle (see Fig. 1). If a transversal (a transverse line) is drawn across a pair of parallel lines (Fig. 2), then corresponding angles are equal and alternate interior angles are equal, as shown. The

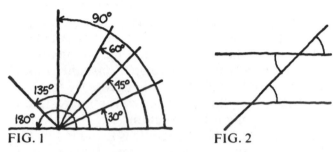

FIG. 1                                    FIG. 2

sum of the angles in any triangle equals 180° (Fig. 3). This can be proved at once by inspecting the diagram shown in Fig. 4. As a direct consequence, we see that the sum of the acute angles

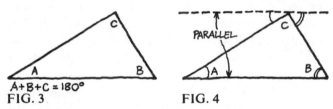

A+B+C = 180°
FIG. 3                          FIG. 4

in a right triangle equals 90° (Fig. 5). Also, in any triangle an exterior angle equals the sum of the opposite interior angles (Fig. 6).

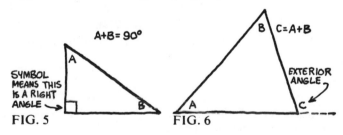

FIG. 5                          FIG. 6

## (b) AREA.

All ideas about area begin in this way: select a unit of length, draw a square whose side is this unit, and define the area of the square to be *one square unit* (Fig. 7). The rectangle shown in Fig. 8 has height 2 and base 3. The internal horizontal

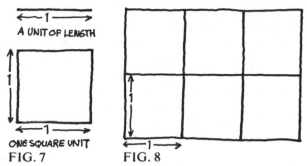

A UNIT OF LENGTH

ONE SQUARE UNIT

FIG. 7    FIG. 8

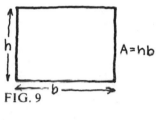

$A = hb$

FIG. 9

and vertical lines divide the rectangle into 6 squares each of area 1, so the area of the rectangle is evidently 6 square units. The fact that $6 = 2 \cdot 3$ suggests that the area $A$ of an arbitrary rectangle of height $h$ and base $b$ should be defined by the formula $A = hb$ (Fig. 9). By Fig. 10, the area of a right triangle of height $h$ and base $b$ is given by

$A = \frac{1}{2}hb$. Since any triangle can be viewed as the

sum or the difference of two right triangles (Fig. 11), for each of which this formula is valid, the formula is valid for all triangles. [Thus, in the triangle on the left in Fig. 11, $b_1$ and $b_2$ are the bases

of the two right triangles and $A = \frac{1}{2}hb_1 + \frac{1}{2}hb_2 =$

$\frac{1}{2}h(b_1 + b_2) = \frac{1}{2}hb$.] In Fig. 12 the two horizontal

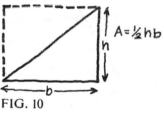

$A = \frac{1}{2}hb$

FIG. 10

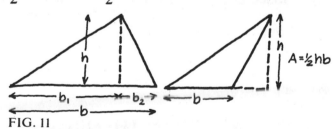

$A = \frac{1}{2}hb$

FIG. 11

lines are understood to be parallel, so all the triangles shown have the same height and the same base, and therefore the same area. We can express this in a different way by saying that if the base of the triangle is held fixed on the lower line and the upper vertex is moved back and forth along the upper line, then the area of the triangle does not change.

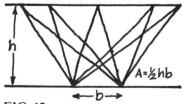

$A = \frac{1}{2}hb$

FIG. 12

## (c) SIMILARITY.

Roughly speaking, two triangles are similar if they have the same shape but different sizes, that is, if one is a magnified version of the other. The precise meaning of similarity for triangles is that their corresponding angles must be equal; and this implies that the ratios of their corresponding sides must also be equal, as shown in Fig. 13. The first of these equations $\left(\dfrac{a}{d}=\dfrac{b}{e}\right)$ can be written in the equivalent form

$$\frac{a}{b}=\frac{d}{e}.$$

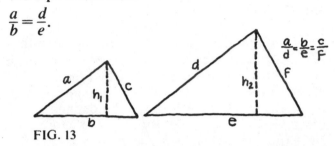

FIG. 13

In words: if two triangles are similar then the ratio of any two sides of one triangle equals the ratio of the corresponding sides of the other. By part (a) above, two triangles will necessarily be similar if two pairs of corresponding angles are equal. Finally, the ratio of the areas of two similar triangles equals the ratio of the squares of any pair of corresponding sides. [To see why this is true, notice in Fig. 13 that if $A_1 = \frac{1}{2}h_1 b$ and $A_2 = \frac{1}{2}h_2 e$ then

$$\frac{A_1}{A_2}=\frac{\frac{1}{2}h_1 b}{\frac{1}{2}h_2 e}=\left(\frac{h_1}{h_2}\right)\left(\frac{b}{e}\right)=\left(\frac{b}{e}\right)\left(\frac{b}{e}\right)=\frac{b^2}{e^2}.\Big]$$

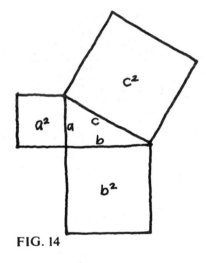

FIG. 14

## (d) PYTHAGOREAN THEOREM.

In its purely geometric form, this famous and indispensable theorem states that in any right triangle, if squares are constructed on the hypotenuse and the two legs, then the square on the hypotenuse equals the sum of the squares on the legs (Fig. 14). In a more algebraic version, it says that the square *of* the hypotenuse equals the sum of the squares *of* the legs. This is illustrated on the left in Fig. 15, and is proved on the right. [The proof is carried out by inserting four replicas of the triangle in the corners of a square of side $a + b$. The first equation on the right says that the area of the large square equals the area of the four triangles plus the area of the small inner square. Why is the inner quadrilateral a square?]

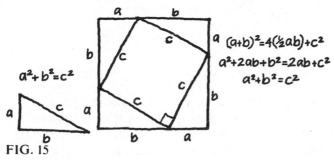

$$(a+b)^2 = 4(\tfrac{1}{2}ab) + c^2$$
$$a^2 + 2ab + b^2 = 2ab + c^2$$
$$a^2 + b^2 = c^2$$

$$a^2 + b^2 = c^2$$

FIG. 15

## 2. CIRCLES

### (a) THE NUMBER $\pi$.

The Greek letter $\pi$ (pronounced 'pie') denotes the ratio of the circumference of a circle to its diameter (Fig. 16):

$$\pi = \frac{c}{2r}, \quad \text{so } c = 2\pi r.$$

The number $\pi$ is known to be irrational. This means that it cannot be expressed exactly as a fraction or as a terminating decimal. However, its numerical value can be calculated in various ways to any specified degree of accuracy. This value is approximately 3.14, or even more accurately, 3.14159. The fraction $\frac{22}{7}$ is also a good approximation, better than 3.14 but not as good as 3.14159.*

### (b) AREA.

Our purpose here is to understand the formula
$$A = \pi r^2$$
for the area $A$ of a circle in terms of its radius $r$.

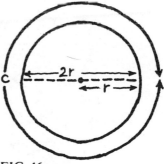

FIG. 16

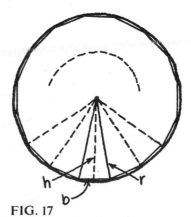

FIG. 17

---

*The student who wishes to learn more about the rich history of this fascinating number should read Isaac Asimov's article "A Piece of Pi," reprinted in his book, *Asimov on Numbers* (Doubleday, 1977). Even better is Petr Beckmann's book, *A History of Pi* (Golem Press, Boulder, Colo., 1971).

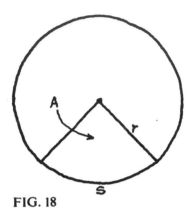

FIG. 18

Suppose that the circle has inscribed in it a regular polygon with a large number of sides (Fig. 17). Each of the small triangles shown in the figure has area $\frac{1}{2}hb$, and the sum of these areas equals the area of the polygon, which is approximately equal to the area of the circle. If $p$ denotes the perimeter of the polygon, we see that

$$A_{\text{polygon}} = \frac{1}{2}hb + \frac{1}{2}hb + \cdots + \frac{1}{2}hb$$

$$= \frac{1}{2}h(b + b + \cdots + b) = \frac{1}{2}hp.$$

Let $c$ be the circumference of the circle. Then, as the number of sides of the polygon increases, $h$ approaches $r$ (this is symbolized by writing $h \to r$, where the arrow means 'approaches'), $p \to c$, and therefore

$$A_{\text{polygon}} = \frac{1}{2}hp \to \frac{1}{2}rc = \frac{1}{2}r(2\pi r) = \pi r^2.$$

This establishes the formula stated above.

To find the area of a sector (Fig. 18), we use the observation that the ratio of this area to the total area of the circle equals the ratio of the intercepted arc $s$ to the complete circumference:

$$\frac{A}{\pi r^2} = \frac{s}{2\pi r}, \text{ so } A = \frac{1}{2}rs.$$

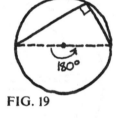

FIG. 19

This fact is easy to remember, since the area of the sector is exactly what it would be if the sector were a triangle with height $r$ and base $s$.

## (c) INSCRIBED ANGLES.

Fig. 19 illustrates the important fact that an angle inscribed in a semicircle is necessarily a right angle. This is most easily understood as a special case of the more general fact (Fig. 20) that an angle inscribed in an arc $ABC$ of a circle always equals one-half of the corresponding central angle. To see why this is true, we begin by considering the simplest special case, in which one side of the inscribed angle passes through the center of the circle (Fig. 21). Here we see that the central angle $x$ is an exterior angle of the indicated isosceles triangle; this central angle therefore equals the sum of the base angles of the

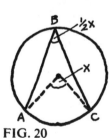

FIG. 20

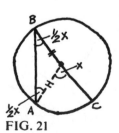

FIG. 21

triangle, so each of these base angles equals $\frac{1}{2}x$.

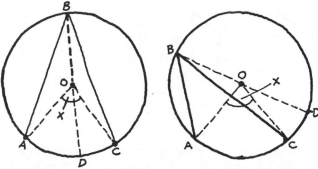

FIG. 22

The other two cases—in which the center of the circle lies inside or outside the inscribed angle (Fig. 22)—can easily be reduced to the case already considered, by drawing the diameter $BD$. [Thus, on the left in Fig. 22, $\angle ABC = \angle ABD + \angle DBC = \frac{1}{2}\angle AOD + \frac{1}{2}\angle DOC = \frac{1}{2}\angle AOC = \frac{1}{2}x$; and on the right, $\angle ABC = \angle ABD - \angle CBD = \frac{1}{2}\angle AOD - \frac{1}{2}\angle COD = \frac{1}{2}\angle AOC = \frac{1}{2}x$.]

A rather surprising conclusion can be drawn from this discussion: if the points $A$ and $C$ are held fixed, and $B$ is moved to various positions on the circle, as shown in Fig. 23, then all of the corresponding inscribed angles are equal to one another.

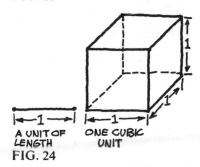

FIG. 23

## 3. CYLINDERS

All ideas about volume begin in this way: select a unit of length, consider a cube whose edge is this unit, and define the volume of this cube to be *one cubic unit* (Fig. 24). The rectangular box shown in Fig. 25 has height 3 and a rectangular base with sides 2 and 4. This box can be divided by horizontal and vertical planes into 3 layers of unit cubes in which each layer contains $2 \cdot 4 = 8$ cubes (in the figure we indicate the horizontal dividing planes). There are clearly $8 \cdot 3 = 24$ unit cubes altogether, so the volume of the box is 24 cubic units. The fact that the volume of this box is the area of the base times the height suggests that the volume $V$ of an arbitrary rectangular box

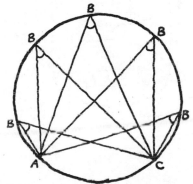

A UNIT OF LENGTH    ONE CUBIC UNIT

FIG. 24

FIG. 25

9

with height $h$ and area of base $B$ should be defined by the formula $V = Bh$ (Fig. 26). Similarly, the

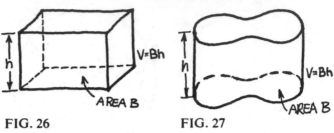

FIG. 26          FIG. 27

volume of any solid with vertical walls and horizontal base and top (Fig. 27) is defined to be the area of the base times the height. In particular, the volume of a cylinder (understood to be a right circular cylinder) with height $h$ and radius of base $r$ (Fig. 28) is $V = \pi r^2 h$, since the area of the base is $\pi r^2$.

Suppose that the top and bottom are removed from a cylinder and that its lateral surface is opened by a vertical cut and unrolled into a rectangle (Fig. 29). It is easy to see that the lateral area of the cylinder is the area of this rectangle, $2\pi rh$, and that the total area is $2\pi rh + 2\pi r^2$.

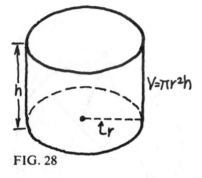

FIG. 28

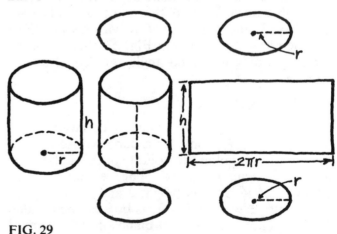

FIG. 29

## 4. CONES

Consider a cone (understood to be a right circular cone) with height $h$, radius of base $r$, and slant height $s$, as shown in Fig. 30. The fundamental facts about cones are the formulas stated in Fig. 30. The volume formula (the volume equals one-third the area of the base times the height, or equivalently, one-third the volume of the circumscribing cylinder) is difficult, and we discuss it in part (b) below. The lateral area formula is easier.

## (a) LATERAL AREA.

The formula for the lateral area is proved in Fig. 31 by cutting the lateral surface of the cone down a generator and unrolling this surface into a sector of a circle.* This formula is easy to remember by thinking of the lateral surface as swept out by revolving a generator about the axis: the lateral area equals the length of this generator multiplied by the distance traveled by its midpoint, $s \cdot 2\pi\left(\frac{1}{2}r\right) = \pi rs$, as suggested in Fig. 30. It is also useful to know that the lateral area of a frustum of a cone (Fig. 32) equals the length of a generator multiplied by the distance traveled by its midpoint. This is proved in the figure.

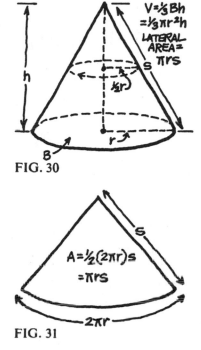

FIG. 30

FIG. 31

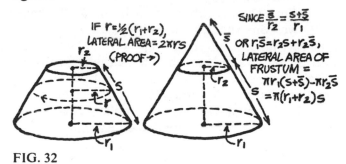

FIG. 32

## (b) VOLUME.

In Fig. 33 we consider the cone shown in Fig. 30. Our purpose is to establish the volume formula $V = \frac{1}{3}Bh$, in particular to understand where the factor $\frac{1}{3}$ comes from. In the base of this cone we inscribe a regular polygon with $n$ sides, where $n$ is some large number (in the figure, $n = 8$). Using this polygon as a base, we construct

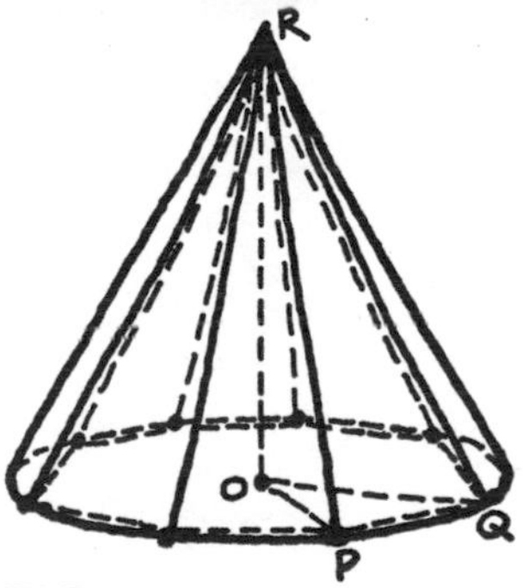

FIG. 33

---

*A *generator* of a cone is a straight line down the side, joining the vertex to a point on the circumference of the base.

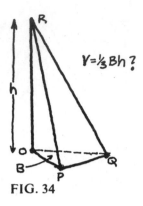

FIG. 34

a pyramid whose vertex is the vertex of the cone. The volume of the cone is the limiting value approached by the volume of the pyramid as $n$ increases. To prove the volume formula for the cone, it therefore suffices to show that the volume of the pyramid is one-third the area of its base times its height. Since the pyramid can be divided into $n$ congruent pyramids of the type shown in Fig. 34, it suffices to show that the volume formula is valid for these special pyramids. This we now do.

On the left in Fig. 35 we show the pyramid in

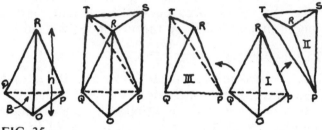

FIG. 35

Fig. 34 in a slightly different position. On the base $OPQ$ we construct a prism with height $h$ and base area $B$ (Fig. 35, center). This prism can be divided into three pyramids as shown on the right in the figure. Pyramids I and II have height $h$ and triangular bases $OPQ$ and $RST$ of equal area, so they have equal volumes. Pyramids II and III have the same height (the distance from $R$ to the plane $PQST$) and triangular bases $PST$ and $PQT$ of equal area, so they also have equal volumes. This argument shows that all three pyramids have the same volume, so the volume of each is one-third the volume $Bh$ of the prism. This establishes the volume formula in Fig. 34, and with it the volume formula for the cone as stated in Fig. 30.*

---

*The argument given here makes use of the fact that two pyramids with the same height and triangular bases with the same area have equal volumes. This fact is proved in Appendix D by means of Cavalieri's Principle as stated in the next section.

# 5. SPHERES

Our purpose is to establish the formulas stated in Fig. 36 for the volume $V$ and surface area $A$ of a sphere. These are profound facts and require ingenious methods.

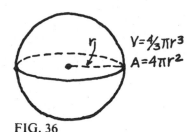

FIG. 36

## (a) CAVALIERI'S PRINCIPLE.

Consider a rectangular solid (Fig. 37, left) consisting of a stack of thin rectangular cards, all with the same dimensions. The shape of this stack can easily be altered without changing its volume, by gently pushing at it horizontally (Fig. 37, right). The volume before is clearly the same as the volume after, since each card in the stack is unchanged except in its position relative to nearby cards. Next, consider two solids with different shapes but the same height (Fig. 38), made up of equal numbers of thin cards. If we assume that each card in one stack has the same face area as the corresponding card in the other stack, regardless of the different shapes of these cards, then it seems reasonable to conclude that the two solids have the same volume. These remarks about the volumes of solids consisting of stacks of thin cards suggest a very powerful principle in the theory of volumes. This principle was first formulated by the Italian mathematician whose name it bears. *Cavalieri's Principle* states that if two solids have the property that every plane parallel to a fixed plane intersects them in cross-sections having equal areas, then the two solids have the same volume (Fig. 39).

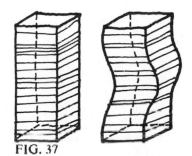

FIG. 37

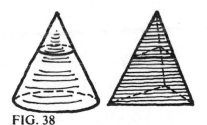

FIG. 38

## (b) THE VOLUME FORMULA.

We find the volume of a sphere of radius $r$ by comparing the sphere with the following solid (Fig. 40): consider a cylinder with base radius $r$

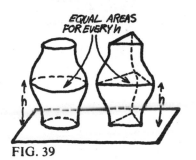

EQUAL AREAS FOR EVERY $h$

FIG. 39

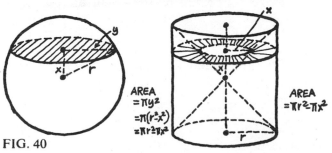

AREA
$= \pi y^2$
$= \pi(r^2-x^2)$
$= \pi r^2 - \pi x^2$

AREA
$= \pi r^2 - \pi x^2$

FIG. 40

and height $2r$; the comparison solid is what remains of this cylinder after the removal of the two cones shown in the figure, that is, it is the cylinder with two conical hollows on the ends. If we calculate the areas of corresponding cross-sections of these solids, as indicated in the figure, we find that they are equal. By Cavalieri's Principle, the solids have equal volumes. The volume $V$ of the sphere, being equal to the volume of the cylinder minus the volumes of the two cones, is therefore given by the formula

$$V = \pi r^2(2r) - 2\left(\frac{1}{3}\pi r^2 \cdot r\right)$$

$$= 2\pi r^3 - \frac{2}{3}\pi r^3$$

$$= \frac{4}{3}\pi r^3.$$

## (c) THE SURFACE AREA FORMULA.

FIG. 41

We find the surface area $A$ of a sphere of radius $r$ by dividing the solid sphere into a large number of small "pyramids." Imagine that the surface of the sphere is divided into a large number of tiny "triangles," as suggested in Fig. 41. These are not actually triangles, since there are no straight line segments on the surface of a sphere. However, being very small, they are nearly triangles. Let each such "triangle" be used as the base of a "pyramid" whose vertex is the center of the sphere. If $a$ is the area of the base of our tiny "pyramid," and $r$ is its height, then its volume $v$ is given by $v = \frac{1}{3}ar$. If we add these equations for all such "pyramids," filling the solid sphere, we see that the volume $V$ and surface area $A$ of the sphere are connected by the equation

$$V = \frac{1}{3}Ar.$$

Since we know that $V = \frac{4}{3}\pi r^3$, we have

$$\frac{4}{3}\pi r^3 = \frac{1}{3}Ar,$$

and therefore
$$A = 4\pi r^2.$$

14

# APPENDIX A. THE MAIN FORMULAS OF GEOMETRY

The formulas stated here express the main content of Sections 1 to 5.

## TRIANGLES (Figs. 42, 43)

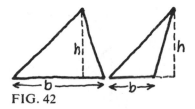
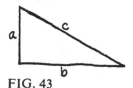

FIG. 42        FIG. 43

area $A = \frac{1}{2}hb$

Pythagorean theorem: $a^2 + b^2 = c^2$

## CIRCLES (Fig. 44)

circumference $c = 2\pi r$
area $A = \pi r^2$

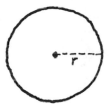

FIG. 44

## CYLINDERS (Fig. 45)

lateral area $A = 2\pi rh$
volume $V = \pi r^2 h$

## CONES (Fig. 46)

lateral area $A = \pi rs$
volume $V = \frac{1}{3}\pi r^2 h$

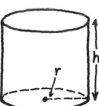
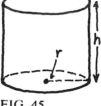

FIG. 45

## SPHERES (Fig. 47)

surface area $A = 4\pi r^2$
volume $V = \frac{4}{3}\pi r^3$

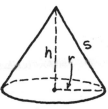

FIG. 46

# APPENDIX B. A NUMBER OF EXERCISES, SOME EASY AND SOME HARD

The harder (and therefore more interesting) exercises are marked with an asterisk (*). The exercises with full solutions given in Appendix C are marked with two asterisks (**).

FIG. 47

# SECTION 1

1. Find the angle $A$ in each of the following figures:

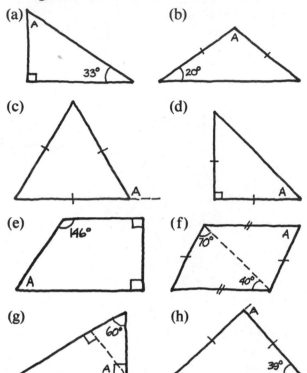

(a)

(b)

(c)

(d)

(e)

(f)

(g)

(h)

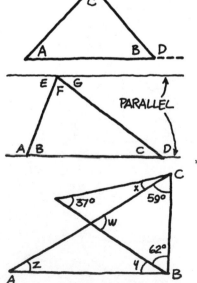

2. In each case find the required angle:
   (a) if $A = 40°$ and $C = 100°$, $D = ?$;
   (b) if $A = 50°$ and $B = 30°$, $C = ?$;
   (c) if $A = 50°$ and $D = 140°$, $C = ?$

3. In each case find the angles not given:
   (a) $A = 150°$, $C = 40°$;
   (b) $B = 60°$, $C = 65°$.

**4. In the figure, $AB \perp BC$ (the symbol $\perp$ means "is perpendicular to"). Find the angles $x$, $y$, $z$, $w$.

5. What is the height of a rectangle whose area is 40 square inches and whose base is 8 inches?

6. If the base of a triangle is 9 inches and its area is 45 square inches, what is its height?

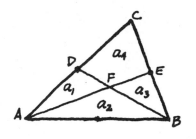

7. The height and base of a triangle are equal and its area is 32 square inches. Find the height and base.

*8. In the triangle $ABC$, $D$ and $E$ are the midpoints of $AC$ and $BC$. The segments $AE$ and $BD$ intersect at $F$. Show that regions $a_2$ and $a_4$ have equal areas.

9. A quarter (a 25¢ piece) is $\frac{3}{4}$ inch in diameter, and when placed 7 feet from the eye will just block out the disc of the moon. If the diameter of the moon is 2160 miles, how far is the moon from the earth?

10. A man 6 feet tall stands at the foot of a flagpole. If the shadow of the man and the pole are 4 feet and 40 feet in length, how tall is the pole?

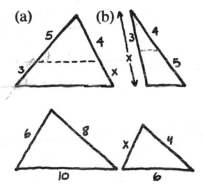

11. In each figure the dotted line is parallel to a side of the triangle. Find $x$.

12. Find $x$ and $y$ in the adjoining similar triangles.

13. The areas of two similar triangles are 25 and 16. If the perimeter of the first is 15, find the perimeter of the second.

14. Two equilateral triangles have sides of 4 inches and 6 inches. What is the ratio of their perimeters? Of their areas? Of their heights?

15. Find $x$ in each of the following right triangles:

    (a)              (b)              (c)

    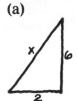        

16. A 20-foot ladder leans against a wall with its foot 6 feet from the wall. A man stands on a rung which is 12 feet from the bottom of the ladder. How far is the man from the wall and from the ground?

17. What is the diagonal of a square whose side is $a$?

18. What is the side of a square whose diagonal is $a$?

17

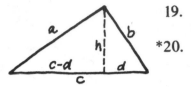

19. Find the area of an equilateral triangle whose side is $a$.

*20. Heron's formula states that the area of a triangle with sides $a$, $b$, $c$ is given by $A = \sqrt{s(s-a)(s-b)(s-c)}$, where the number $s = \frac{1}{2}(a+b+c)$ is called the semiperimeter. Prove this formula by verifying the following steps:

(a) $A = \frac{1}{2}hc$;

(b) $a^2 = b^2 + c^2 - 2cd$;

(c) $d = \dfrac{b^2 + c^2 - a^2}{2c}$;

(d) $h^2 = b^2 - d^2 = b^2 - \dfrac{(b^2 + c^2 - a^2)^2}{4c^2}$

$\quad = \dfrac{(a+b+c)(b+c-a)(a+b-c)(a-b+c)}{4c^2}$

$\quad = \dfrac{2s(2s-2a)(2s-2c)(2s-2b)}{4c^2}$

$\quad = \dfrac{4s(s-a)(s-b)(s-c)}{c^2}$;

(e) $h = \dfrac{2}{c}\sqrt{s(s-a)(s-b)(s-c)}$;

(f) $A = \sqrt{s(s-a)(s-b)(s-c)}$.

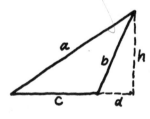

These steps require changes for a triangle of the shape indicated at the left. Provide these changes, and in this way show that the area formula is valid without any restrictions at all.

21. Use Heron's formula and one other method to find the area of a right triangle with sides 3, 4, 5.

22. Apply Heron's formula to verify the result of Exercise 19.

23. Find the hypotenuse of a right triangle whose legs are (a) 3, 4; (b) 5, 12; (c) 6, 8; (d) 7, 24; (e) 8, 15.

24. The hypotenuse of a right triangle is 15 and one leg is 12. What is the other leg?

**25. If $E$ is any interior point of the indicated rectangle, show that $a^2 + c^2 = b^2 + d^2$.

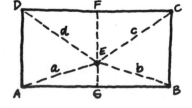

26. In the adjoining figure, find $a$, $b$, $c$, $d$, $e$.

*27. Show that in any parallelogram the sum of the squares of the diagonals equals the sum of the squares of the four sides.

28. A side of one square equals a diagonal of a second square. Find the ratio of the area of the larger square to that of the smaller square.

29. A side of one equilateral triangle equals the height of a second equilateral triangle. Find the ratio of the perimeter of the larger triangle to that of the smaller.

30. Show that in a 30°-60° right triangle the altitude on the hypotenuse divides the hypotenuse into segments whose lengths have the ratio 1/3.

## SECTION 2

1. The diameter of one circle equals the radius of a second circle. Find the ratio of their areas.

2. The ratio of the areas of two circles of radii $R$ and $r$ is 2/1. What is the ratio $R/r$ of their radii?

3. Find the area of a sector of a circle of radius 10 whose central angle is (a) 60°; (b) 90°; (c) 48°.

4. Two concentric circles have circumferences $30\pi$ and $40\pi$. Find the area of the ring-shaped region between them.

5. Show that the area of the ring-shaped region between two concentric circles equals the area of a circle whose diameter is a chord of the outer circle which is tangent to the inner circle.

*6. Find the area of the shaded part of each figure:

(a)    (b)    (c)

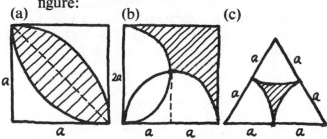

7. A goat is tied to the corner of a shed 12 feet long and 10 feet wide. If the rope is 15 feet long, over how many square feet can the goat graze?

8. The earth is approximately 93 million miles from the sun. Assuming that the earth's orbit is circular, approximately how far does the earth move along its orbit in each second? Use the approximation $\pi = \frac{22}{7}$.

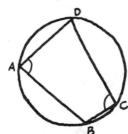

9. Consider two circles with the second internally tangent to the first at a point $A$ and also passing through the center of the first. Show that every chord of the first circle which has $A$ as an endpoint is bisected by the second circle.

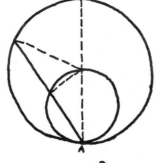

10. If $ABCD$ is a quadrilateral inscribed in a circle, show that the opposite angles $A$ and $C$ are supplementary ($A + C = 180°$).

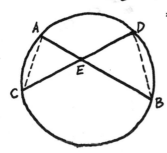

**11. $AB$ and $CD$ are two chords in a circle, and they intersect at a point $E$. Show that the product of the segments of one chord equals the product of the segments of the other chord, that is, that $AE \cdot BE \doteq CE \cdot DE$. Hint: use similar triangles.

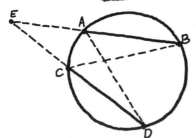

12. If $AB$ and $CD$ are two chords of a circle that have been extended to intersect at an external point $E$, show that $AE \cdot BE = CE \cdot DE$.

# SECTION 3

1. Find the volume and the total surface area of a rectangular box with edges 4, 5 and 6 feet.

2. Find the length of the diagonal joining two opposite vertices of the box in the preceding exercise.

3. What is the volume of a cube whose total surface area is 150 square inches?

4. A cylinder is 7 inches high and the radius of its base is 4 inches. Use the approximation $\pi = \frac{22}{7}$ and calculate (a) its volume; (b) its lateral area; (c) its total area.

**5. If the radius of the base of a cylinder is doubled and its height is tripled, by what number is the volume multipled?

6. Find the volume of a cylinder if the radius of its base is one-third its height $h$.

7. In a certain cylinder the lateral area is half the total area. How is the height $h$ related to the radius $r$ of the base?

# SECTION 4

1. Find the volume of a cone 28 feet high whose base has diameter 12 feet. Use the approximation $\pi = \frac{22}{7}$.

2. Find the height of a cone whose volume is 484 cubic inches and whose base has radius 7 inches. Use the approximation $\pi = \frac{22}{7}$.

**3. The height of a cone equals the radius $r$ of its base. Show that the volume $V$ is given by the formula $V = \frac{\sqrt{2}}{6} rA$ where $A$ is the lateral area.

*4. The height of a cone is $h$. A plane parallel to the base intersects the axis at a certain point. How far from the vertex must this point be if the plane divides the lateral area into two equal parts? If the plane divides the volume into two equal parts?

5. The height of a cone is $h$ and the radius of its base is $r$. If $r$ is halved, how must $h$ be changed to keep the volume unchanged?

6. A plane parallel to the base of a cone bisects the axis. What is the ratio of the volume of the original cone to the volume of the frustum formed in this way?

7. The Great Pyramid of Egypt originally had a square base 755 feet on a side and was 481 feet high. Compute its volume.

8. The adjoining cube consists of six identical pyramids. Find the volume of one of these pyramids in two different ways.

9. A conical tent is made by using a semicircular piece of canvas of radius 8 feet. Find the height of the tent and the number of cubic feet of air inside.

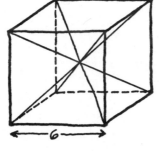

# SECTION 5

1. Find the volume of a sphere whose diameter is 6.

2. Find the radius of a sphere whose volume is $2304\pi$.

3. Find the area of a sphere if the circumference of a great circle is $40\pi$. (A great circle is the intersection of the surface of a sphere with a plane through its center.)

4. Find the radius of a sphere whose area is $36\pi$.

5. The radius of the earth is approximately 4000 miles and the area of the continental United States is approximately 3,000,000 square miles. What percentage of the total area of the earth does this area represent? Use the approximation $\pi = \frac{22}{7}$.

6. A cylinder is circumscribed about a sphere. Find the ratio of the volume of the cylinder to the volume of the sphere.

7. A cylinder is circumscribed about a sphere. Show that the area of the sphere equals the lateral area of the cylinder.

*8. If the radius of a certain sphere is increased by 6, its area is multiplied by 9. Find the radius of the sphere.

9. A cylinder with height 6 and radius 4 is inscribed in a sphere. Find the area and volume of the sphere.

10. A cylinder is circumscribed about a sphere. A cone has the same base and height as the cylinder. Find the ratio of the total area of the cylinder to that of the sphere and of the cone.

*11. An equilateral triangle and a square are inscribed in a circle, with a side of the triangle being parallel to a side of the square. The entire figure is revolved about that altitude of the triangle which is perpendicular to a side of the square. Find the ratio of the area of the sphere to the total area of the cylinder, and the ratio of the total area of the cylinder to the total area of the cone.

**12. A sphere is inscribed in a cone. The slant height of the cone equals the diameter of its base. If the height of the cone is 9, find the area of the sphere.

*13. The ratio of the volumes of two spheres is 27/343 and the sum of their radii is 10. What is the radius of the smaller sphere?

14. A sphere is circumscribed about a cube. Find the ratio of the volume of the cube to the volume of the sphere.

15. A cylinder is circumscribed about a sphere. Show that the ratio of their volumes equals the ratio of their total areas.

16. Two spheres of radii 3 inches and 5 inches rest on a table and touch one another. How far apart are the points at which they touch the table?

*17. Use Cavalieri's Principle to find the volume of a spherical segment of one base and thickness $h$ if the radius of the sphere is $r$.

*18. In the preceding exercise, find the volume of the spherical sector (the solid shown on the right, resembling a filled ice cream cone). Use this result, and a comparison of areas and volumes, to show that the area of the curved surface on top of the sector is $2\pi rh$.

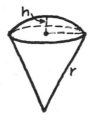

*19. A spherical ring is the solid that remains after removing from a solid sphere of radius $r$ a cylindrical boring whose axis passes through the center of the sphere. If $h$ is the height of the ring, use the result of Exercise 17 to show that the volume of the ring is $\frac{\pi h^3}{6}$. (Notice how remarkable it is that this volume depends only on $h$, and not on the radius $r$ of the sphere.)

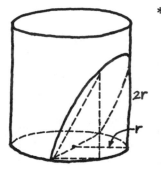

*20. A cylindrical wedge is the solid cut from a cylinder by a tilted plane passing through a diameter of the base. Apply Cavalieri's Principle to find the volume of such a wedge if its height is $2r$, where $r$ is the radius of the base and the height of the cylinder is $> 2r$. (Use as a comparison solid a rectangular box having edges $r$, $r$, $2r$ with two square pyramids removed, where the pyramids have the square ends of the box as bases and common vertex at the center of the box. Stand the box on one of its square ends and place the wedge so that the bounding diameter of its base is vertical.)

# APPENDIX C. THE ANSWERS TO THE EXERCISES, WITH FULL SOLUTIONS FOR A SELECTED FEW

## SECTION 1

**1.** (a) 57°; (b) 140°; (c) 120°; (d) 45°; (e) 34°; (f) 70°; (g) 60°; (h) 78°.

**2.** (a) $D = 140°$; (b) $C = 100°$; (c) $C = 90°$.

**3.** (a) $B = 30°$, $D = 140°$, $E = 30°$, $F = 110°$, $G = 40°$; (b) $A = 120°$, $D = 115°$, $E = 60°$, $F = 55°$, $G = 65°$.

**4.** Since the sum of the angles of a triangle equals 180°, we have $(x + 59°) + 62° + 37° = 180°$, and therefore $x = 180° - (59° + 62° + 37°) = 180° - 158° = 22°$. Since a right angle equals 90°, $y + 62° = 90°$ and therefore $y = 28°$. Since the sum of the acute angles in a right triangle equals 90°, $z + 59° = 90°$ and $z = 31°$. Finally, $w + 59° + 62° = 180°$, so $w = 180° - 121° = 59°$.

**5.** 5 inches.   **6.** 10 inches.   **7.** $h = b = 8$ inches.

**9.** 241,920 miles. **10.** 60 feet. **11.** (a) $\frac{12}{5}$; (b) $\frac{27}{4}$.

**12.** $x = \frac{18}{5}$, $y = \frac{24}{5}$. **13.** 12. **14.** 2/3, 4/9, 2/3.

**15.** (a) $2\sqrt{10}$; (b) $4\sqrt{2}$; (c) $4\sqrt{3}$.

**16.** $\frac{12}{5}$ feet from the wall, $\frac{6}{5}\sqrt{91}$ feet from the

ground. **17.** $\sqrt{2}a$. **18.** $\frac{1}{2}\sqrt{2}a$. **19.** $\frac{1}{4}\sqrt{3}a^2$.

**21.** 6. **23.** (a) 5; (b) 13; (c) 10; (d) 25; (e) 17.
**24.** 9. **25.** By the Pythagorean theorem, we
see that $a^2 + c^2 = AG^2 + GE^2 + CF^2 + FE^2$
and $b^2 + d^2 = BG^2 + GE^2 + DF^2 + FE^2$. But
these expressions are equal, because $AG = DF$
and $BG = CF$.
**26.** $a = \sqrt{2}$, $b = \sqrt{3}$, $c = \sqrt{4}$, $d = \sqrt{5}$, $e = \sqrt{6}$.

**28.** 2. **29.** $\frac{2}{3}\sqrt{3}$.

## SECTION 2

**1.** 1/4. **2.** $R/r = \sqrt{2}$.

**3.** (a) $16\frac{2}{3}\pi$; (b) $25\pi$; (c) $13\frac{1}{3}\pi$. **4.** $175\pi$.

**6.** (a) $\frac{1}{2}a^2(\pi - 2)$; (b) $\frac{1}{2}a^2(6 - \pi)$;

(c) $\frac{1}{2}a^2(2\sqrt{3} - \pi)$. **7.** $177\frac{1}{4}\pi$ square feet.

**8.** Approximately 18 miles. **11.** The angles
$ACD$ and $ABD$ are equal, because they are
inscribed in the same arc of the circle. The
angles $CAB$ and $CDB$ are equal for the same
reason. The angles $AEC$ and $DEB$ are evidently
equal. The triangles $ACE$ and $DBE$ are there-
fore similar, and consequently $\frac{AE}{DE} = \frac{CE}{BE}$ or

$AE \cdot BE = CE \cdot DE$.

## SECTION 3

**1.** Volume = 120 cubic feet, area = 148 square
feet. **2.** $\sqrt{77}$ feet. **3.** 125 cubic inches.
**4.** (a) 352 cubic inches; (b) 176 square inches;

(c) $276\frac{4}{7}$ square inches. **5.** If the original radius

and height are $r$ and $h$, then the new radius and
height are $2r$ and $3h$. The original volume is

$\pi r^2 h$ and the new volume is $\pi(2r)^2(3h) = 12\pi r^2 h$, so the new volume is 12 times the original volume. **6.** $\frac{\pi}{9}h^3$. **7.** $h = r$.

## SECTION 4

**1.** 1056 cubic feet. **2.** $9\frac{3}{7}$ inches.

**3.** Since $r = h$, $A = \pi r \sqrt{r^2 + h^2} = \pi r \sqrt{2h^2} = \pi\sqrt{2}rh$, and $V = \frac{1}{3}\pi r^2 h = \frac{1}{3}\pi r(rh) = \frac{1}{3}\pi r\left(\frac{A}{\pi\sqrt{2}}\right)$
$= \frac{rA}{3\sqrt{2}} = \frac{rA}{3\sqrt{2}} \cdot \frac{\sqrt{2}}{\sqrt{2}} = \frac{\sqrt{2}}{6}rA$.

**4.** $\frac{1}{\sqrt{2}}h$ for the lateral area, $\frac{1}{\sqrt[3]{2}}h$ for the volume.

**5.** $h$ must be multiplied by 4. **6.** 8/7.

**7.** $91{,}394{,}008\frac{1}{3}$ cubic feet. **8.** 36.

**9.** Height $= 4\sqrt{3}$ feet, volume $= \frac{64}{3}\sqrt{3}\,\pi$ cubic feet.

## SECTION 5

**1.** $36\pi$. **2.** 12. **3.** $1600\pi$. **4.** 3. **5.** About 1.5%.

**6.** 3/2. **8.** 3. **9.** $A = 100\pi$, $V = 500\frac{\pi}{3}$.

**10.** $\frac{A_{\text{cyl}}}{A_{\text{sph}}} = \frac{3}{2}$, $\frac{A_{\text{cyl}}}{A_{\text{cone}}} = \frac{6}{\sqrt{5} + 1}$.

**11.** $\frac{A_{\text{sph}}}{A_{\text{cyl}}} = \frac{A_{\text{cyl}}}{A_{\text{cone}}} = \frac{4}{3}$.

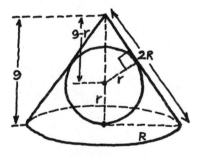

**12.** Draw a good picture of the situation, as shown. If $r$ is the radius of the sphere and $R$ is the radius of the base of the cone, then by using similar triangles we see that $\frac{9 - r}{r} = \frac{2R}{R} = 2$, so $\frac{9}{r} - 1 = 2$, $\frac{9}{r} = 3$, $r = 3$, and therefore $A = 4\pi r^2 = 4\pi 9 = 36\pi$. The similar triangles mentioned are two right triangles with an acute angle in common, and the equation first written involves the ratio of the hypotenuse to the shorter leg in each. **13.** 3. **14.** $2\sqrt{3}/3\pi$. **16.** $2\sqrt{15}$ inches.

**17.** $V = \pi h^2\left(r - \frac{h}{3}\right)$. **18.** $V = \frac{2}{3}\pi r^2 h$. **20.** $\frac{4}{3}r^3$.

# APPENDIX D. TRIANGULAR PYRAMIDS WITH EQUAL HEIGHTS AND BASES

Our purpose is to fill a crucial gap in the argument of Section 4 (see the footnote in that section).

Consider a pyramid with vertex $V$ and triangular base $ABC$ in a horizontal plane (Fig. 48). Its height $h$ is the perpendicular distance from $V$ to the plane of $ABC$. Let the pyramid be cut by a second horizontal plane whose distance above the first plane is $k$, where $k < h$. This second plane intersects the pyramid in a triangle $A'B'C'$ which is similar to $ABC$. We assert that the areas of these triangles are related by the equation

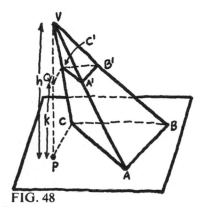

FIG. 48

$$\text{area } A'B'C' = \left(\frac{h-k}{h}\right)^2 \cdot \text{area } ABC.$$

[Proof: By similar triangles,

$$\frac{VC'}{VC} = \frac{VQ}{VP} = \frac{h-k}{h} \quad \text{and} \quad \frac{A'C'}{AC} = \frac{VC'}{VC}$$

$$= \frac{VQ}{VP} = \frac{h-k}{h},$$

so

$$\frac{\text{area } A'B'C'}{\text{area } ABC} = \left(\frac{A'C'}{AC}\right)^2 = \left(\frac{h-k}{h}\right)^2 \cdot \bigg]$$

Next, consider two pyramids with triangular bases in the same horizontal plane (Fig. 49). If they have the same base area $B$ and the same height $h$, then they have the same volume. [Proof: By the above paragraph, horizontal cross-sections at height $k$ have the same area $B_k$, where

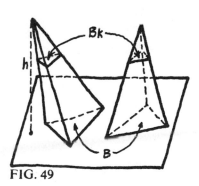

FIG. 49

$$B_k = \left(\frac{h-k}{h}\right)^2 B.$$

Cavalieri's Principle now implies that the pyramids have the same volume.]

# APPENDIX E. CEVA'S THEOREM

It is not difficult to see that the angle bisectors of any triangle are concurrent, in the sense that they all pass through a single point (Fig. 50). Many facts about concurrent lines associated with triangles can be understood by means of a remarkable theorem discovered by the 17th century Italian mathematician Giovanni Ceva.

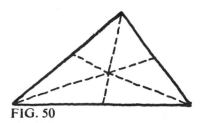

FIG. 50

In order to formulate this theorem it is convenient to introduce the following term. A line segment joining a vertex of a triangle to a point on the opposite side is called a *cevian*. In each triangle of Fig. 51 we show three cevians $AX$, $BY$, $CZ$; on the left these cevians are concurrent

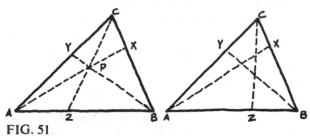

FIG. 51

at a point $P$, but on the right they are not concurrent. Ceva's theorem gives a criterion for concurrence in terms of the lengths of the six segments into which three such cevians divide the sides of the triangle.

## CEVA'S THEOREM

Three cevians $AX$, $BY$ and $CZ$ of a triangle $ABC$ are concurrent if and only if

$$\frac{AZ}{BZ} \cdot \frac{BX}{CX} \cdot \frac{CY}{AY} = 1.$$

Proof. We first assume that the cevians are concurrent at $P$, as shown on the left in Fig. 51. For convenience we denote the area of a triangle $ABC$ by the symbol $(ABC)$. Since the areas of triangles with equal heights are proportional to their bases, we have

$$\frac{AZ}{BZ} = \frac{(ACZ)}{(BCZ)} = \frac{(APZ)}{(BPZ)} = \frac{(ACZ) - (APZ)}{(BCZ) - (BPZ)}$$

$$= \frac{(APC)}{(BPC)}. \quad *$$

In the same way,

$$\frac{BX}{CX} = \frac{(APB)}{(APC)} \quad \text{and} \quad \frac{CY}{AY} = \frac{(BPC)}{(APB)}.$$

---

*In the middle step of this calculation we are essentially saying that if four numbers $a$, $b$, $c$, $d$ satisfy the equation $\frac{a}{b} = \frac{c}{d}$, then $\frac{a}{b} = \frac{c}{d} = \frac{a-c}{b-d}$.

This is easy to verify by cross-multiplication.

By multiplying these three equations together, we obtain our conclusion,

$$\frac{AZ}{BZ} \cdot \frac{BX}{CX} \cdot \frac{CY}{AY} = \frac{(APC)}{(BPC)} \cdot \frac{(APB)}{(APC)} \cdot \frac{(BPC)}{(APB)} = 1.$$

To prove the other half of the theorem, we assume that

$$\frac{AZ}{BZ} \cdot \frac{BX}{CX} \cdot \frac{CY}{AY} = 1.$$

Let $AX$ and $BY$ intersect at $P$ (Fig. 52) and let $CP$ extended intersect $AB$ at $Z'$. Then by the first part of the proof we know that

$$\frac{AZ'}{BZ'} \cdot \frac{BX}{CX} \cdot \frac{CY}{AY} = 1.$$

These two equations imply that

$$\frac{AZ'}{BZ'} = \frac{AZ}{BZ},$$

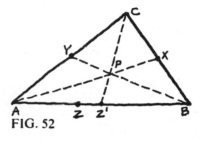

FIG. 52

so $Z' = Z$. It follows that $CZ'$ and $CZ$ are the same segment, so $AX$, $BY$, $CZ$ are concurrent.

Ceva's theorem has a number of immediate consequences.

1. If $X$, $Y$, $Z$ are the midpoints of the sides of the triangle, then by Ceva's theorem the cevians are easily seen to be concurrent. In this case the cevians are usually called *medians*, so we know that the three medians of any triangle intersect at a single point.

2. Ceva's theorem also implies that the three altitudes from the vertices to the opposite sides are concurrent. The proof of this requires a little trigonometry (Fig. 53): if the sides of the triangle are denoted by $a$, $b$, $c$, then $AZ = b \cos A$, $BZ = a \cos B$, $BX = c \cos B$, $CX = b \cos C$, $CY = a \cos C$ and $AY = c \cos A$, so

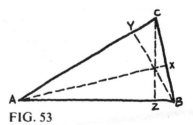

FIG. 53

$$\frac{AZ}{BZ} \cdot \frac{BX}{CX} \cdot \frac{CY}{AY} = \frac{b \cos A}{a \cos B} \cdot \frac{c \cos B}{b \cos C} \cdot \frac{a \cos C}{c \cos A} = 1.$$

3. In the 19th century a French mathematician named Joseph Gergonne proved the following: if a circle is inscribed in a triangle $ABC$ (Fig. 54), and if $X$, $Y$, $Z$ are the points where the circle is tangent to the sides of the triangle, then $AX$, $BY$ and $CZ$ are concurrent. Why is this true?

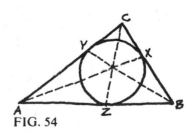

FIG. 54

29

# APPENDIX F.
# BRAHMAGUPTA'S FORMULA

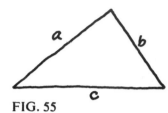

FIG. 55

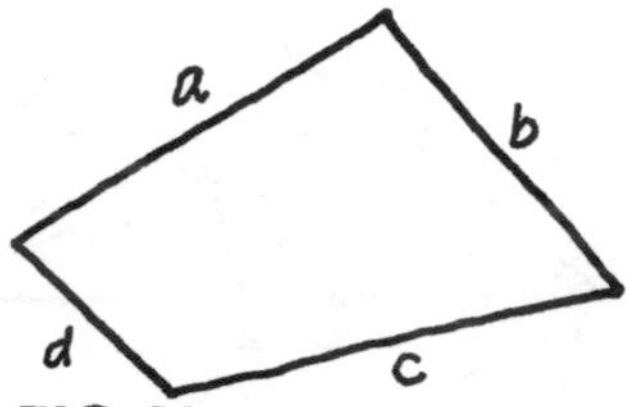

In Exercise 20 of Section 1 we ask the reader to prove Heron's formula for the area $A$ of a triangle with sides $a$, $b$, $c$ (Fig. 55):

$$A = \sqrt{s(s-a)(s-b)(s-c)},$$

where $s = \frac{1}{2}(a + b + c)$ is the semiperimeter of

the triangle. The presence of the factor $s$ under the radical suggests that this formula might be a special case of a more general formula

$$A = \sqrt{(s-a)(s-b)(s-c)(s-d)} \qquad (*)$$

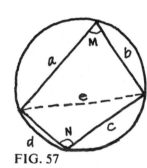

FIG. 56

giving the area of a quadrilateral with sides $a$, $b$, $c$, $d$, where $s = \frac{1}{2}(a + b + c + d)$ is the semiperimeter of the quadrilateral (Fig. 56). After all, if the side $d$ shrinks to zero, then the quadrilateral becomes a triangle and formula (*) collapses to Heron's formula, which we know is correct. Unfortunately this conjecture is false (can you see at once why it cannot be true?). However, a modified version is true: if the quadrilateral is inscribed in a circle (Fig. 57), then formula (*) is valid. Under these circumstances (*) is called Brahmagupta's formula, after the 7th century Indian mathematician who discovered it.

FIG. 57

The proof we give makes use of trigonometry. We begin by inserting the diagonal $e$ in the quadrilateral of Fig. 57, and also by labeling the opposite angles $M$ and $N$. By Exercise 10 in Section 2, $M + N = 180°$, so

$$\cos N = -\cos M \quad \text{and} \quad \sin N = \sin M.$$

By the law of cosines,

$$a^2 + b^2 - 2ab \cos M = e^2 = c^2 + d^2 - 2cd \cos N,$$

so

$$2(ab + cd)\cos M = a^2 + b^2 - c^2 - d^2. \qquad (**)$$

Since the area $A$ of the quadrilateral is given by

$$A = \frac{1}{2}ab \sin M + \frac{1}{2}cd \sin N = \frac{1}{2}(ab + cd)\sin M,$$

we also have

$$2(ab + cd)\sin M = 4A. \qquad (***)$$

By squaring and adding equations (**) and (***),

30

and using the identity $\sin^2 M + \cos^2 M = 1$, we obtain

$$4(ab + cd)^2 = (a^2 + b^2 - c^2 - d^2)^2 + 16A^2,$$

so

$$16A^2 = (2ab + 2cd)^2 - (a^2 + b^2 - c^2 - d^2)^2.$$

By repeatedly factoring differences of two squares in accordance with the identity $x^2 - y^2 = (x + y) \times (x - y)$, we obtain

$$\begin{aligned}
16A^2 &= [2ab + 2cd + a^2 + b^2 - c^2 - d^2] \\
&\quad \times [2ab + 2cd - a^2 - b^2 + c^2 + d^2] \\
&= [(a + b)^2 - (c - d)^2] \\
&\quad \times [(c + d)^2 - (a - b)^2] \\
&= [a + b + c - d][a + b - c + d] \\
&\quad \times [c + d + a - b][c + d - a + b] \\
&= [2s - 2d][2s - 2c][2s - 2b][2s - 2a]
\end{aligned}$$

or

$$A^2 = (s - a)(s - b)(s - c)(s - d),$$

which proves Brahmagupta's formula.

# CHAPTER 2
# ALGEBRA

"By relieving the brain of all unnecessary work, a good notation sets it free to concentrate on more advanced problems, and in effect increases the mental power of the race. . . . By the aid of symbolism, we can make transitions almost mechanically by the eye, which otherwise would call into play the higher faculties of the brain. It is a profoundly erroneous truism, that we should cultivate the habit of thinking of what we are doing. The precise opposite is the case. Civilization advances by extending the number of important operations which we can perform without thinking about them."

—Alfred North Whitehead

"The more I work and practice, the luckier I seem to get."

—Gary Player
(professional golfer)

# INTRODUCTION

Most American schools are now in full or partial retreat from the ill-fated educational experiment known as "the New Math." The purpose of the New Math was to revitalize American mathematics education in the wake of the Russian space achievements in the 1950's. However, its method was to emphasize form over substance to the detriment of both. The damage was especially heavy in the field of algebra, as more and more students came along who had heard of the commutative law but did not know the multiplication table. The result was two decades of steady decline in the teaching (and learning) of algebra, but things are now changing and substance is on its way back.

Algebra, like grammar, has very little pizzazz. A few hardy souls find grammar interesting, but for most people it is dull and cannot be made otherwise. And so it is with algebra. Even though there are some teachers and students who do find grammar interesting, and it can be expected that algebra will likewise have its small band of enthusiasts, I make no effort here to con the student into believing that algebra is useful or exciting for everyone in everyday life. It isn't. Its importance lies in the student's future—and even then only for some students—as essential preparation for the serious study of science, engineering, economics, or some more advanced type of mathematics. Since the aim of this book is to be especially helpful to these groups of students, I have made a particular effort to trim off the fat, to make certain that every page counts, and to follow the advice of the old maxim for political speakers: "If you must be dull, at least have the good sense to be brief."

# 1. BASICS

## 1.1 THE REAL LINE

The basic numbers used in algebra are the *real numbers*.* The system of all real numbers is quite difficult to define in a satisfactory logical way. We do not attempt this task. From the point of view of the student a description leading to a solid intuitive grasp is better than a definition, and we build this description as follows.

The real number system contains several types of numbers of particular importance:

(a) the *positive integers* (or *natural numbers*) 1, 2, 3, 4, 5, . . . ;

(b) the *integers* . . . , −3, −2, −1, 0, 1, 2, 3, . . . ;

(c) the *rational numbers,* which are those real numbers that can be written as fractions (or quotients of integers), such as $\frac{3}{4}$, $-\frac{7}{3}$, 5, 0, −3, 2.614, $7\frac{2}{3}$.

A real number that is not rational is said to be *irrational.* For example, the numbers

$$\sqrt{2}, \sqrt{3}, \sqrt{5}, \quad \text{and} \quad \pi$$

are irrational, though these facts are not easy to prove. (We remind the student that for any positive number $a$, the symbol $\sqrt{a}$ always means its positive square root. Accordingly, $\sqrt{4}$ is equal to 2 and not −2, even though $(-2)^2 = 4$.)

The numbers described in (a), (b) and (c) above form increasingly inclusive subsystems of the system of all real numbers, as suggested on the left in Fig. 1. We can also separate the real num-

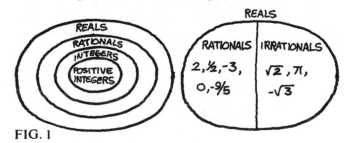

FIG. 1

---

bers into the rationals and the irrationals, as shown on the right in the figure.

**Example.** Give a precise classification of each of the following numbers (that is, if irrational say so, and if rational, state the smallest of the above subsystems to which the number belongs): $-\sqrt{3}$, $\frac{7}{2}$, $-\frac{6}{3}$, $(-2)(-4)$, $\frac{\pi}{5}$, $-17$, $-\frac{95}{3}$.

**Solution.** The numbers $-\sqrt{3}$ and $\frac{\pi}{5}$ are irrational; $\frac{7}{2}$ and $-\frac{95}{3}$ are rationals; $-\frac{6}{3} = -2$ and $-17$ are integers; and $(-2)(-4) = 8$ is a positive integer.

However, the most useful picture of the real number system is obtained by representing these numbers graphically by points on a horizontal straight line, as follows.

This representation (Fig. 2) begins with the

FIG. 2

choice of an arbitrary point as the origin or zero point, and another arbitrary point to the right of it as the point 1. The distance between these two points (the unit distance) then serves as a measuring scale by means of which we can assign a point on the line to every positive and negative integer, and also to every rational number, as indicated in the figure. We have also plotted the three irrational numbers $\sqrt{2}$, $\sqrt{3}$, and $\pi$, whose decimal expansions are infinite and nonrepeating:

$$\sqrt{2} = 1.414\ldots, \qquad \sqrt{3} = 1.732\ldots,$$
$$\pi = 3.14159\ldots.$$

We can now describe the real numbers as all those numbers that correspond to all points on the line, as shown in Fig. 2. This line itself is called the *real line*, or sometimes the *number line*.

In essence, the real numbers are the numbers used in counting and measuring, which are the basic quantitative activities of the human mind. The system of real numbers is evidently an intellectual tool of very great importance, and algebra is the language we use in working with this tool.

## EXERCISE

1.  Give a precise classification of each of the following numbers: $\sqrt{9}$, $-\frac{2}{3}$, $\frac{51}{3}$, $-10$, $-\frac{\pi}{3}$, $\frac{\sqrt{5}}{2}$, $-\sqrt{4}$, $\frac{5}{1234}$.

# 1.2 NOTATION AND THE SIMPLEST RULES OF PROCEDURE

Before beginning our brief survey of algebra, we touch lightly on a few miscellaneous preliminary ideas.

(1) Arithmetic deals with calculations involving particular numbers, algebra with calculations involving general numbers. Thus,

$$(5+2)(5-2) = 7 \cdot 3 = 21 = 25 - 4 = 5^2 - 2^2$$

is a particular fact of no special interest, but

$$(a+b)(a-b) = a^2 - b^2$$

is a universal fact of considerable value. One of the distinctive features of algebra is that it achieves the generality associated with universal facts by the notational device of using letters to represent unspecified numbers. Further, it is customary to represent constants by letters at the beginning of the alphabet ($a$, $b$, $c$, etc.) and to represent unknown or variable quantities by letters at the end of the alphabet ($x$, $y$, $z$, etc.). However, this is not a hard-and-fast rule, and the statements

$$(a+b)(a-b) = a^2 - b^2 \quad \text{and} \quad (x+y)(x-y) = x^2 - y^2$$

have exactly the same meaning.

(2) Division by zero is not permitted, because no definite meaning can be attached to this operation. To understand this, notice that $\frac{a}{0} = b$ must (if it means anything) mean the same as $a = 0 \cdot b$. However, if $a \neq 0$ then $a = 0 \cdot b$ is not true for any $b$; and if $a = 0$ then $a = 0 \cdot b$ is true for every $b$. This tells us that $\frac{a}{0}$ cannot be assigned a definite value—and therefore has no meaning.

(3) Parentheses ( ) and brackets [ ] and braces { } are symbols of grouping that mean the quan-

tity between them is to be treated as a single number, as in

$$a - [3a - (2a + b)] = a - [3a - 2a - b]$$
$$= a - 3a + 2a + b = b.$$

This calculation shows that such symbols can be removed from the inside out, changing signs throughout when the grouping symbol is preceded by a minus sign. Grouping symbols are also used to prevent ambiguity and misunderstanding, as follows:

$1 - (a + b)$   is not the same as   $1 - a + b$;

$\frac{1}{2}(a + b)$   is not the same as   $\frac{1}{2}a + b$;

$1/(a + b)$ means $\dfrac{1}{a + b}$,   while

$1/a + b$ means $\dfrac{1}{a} + b$.

(4) The *commutative* and *associative laws* for addition are

$$a + b = b + a \quad \text{and} \quad a + (b + c) = (a + b) + c;$$

and for multiplication are

$$ab = ba \quad \text{and} \quad a(bc) = (ab)c.$$

These are almost automatic for most people from their experience with arithmetic, and we say no more about them. It is different with the various forms of the *distributive law:*

$$a(b + c) = ab + ac, \qquad (a + b)c = ac + bc,$$
$$a(b - c) = ab - ac, \qquad (a - b)c = ac - bc.$$

These are useful for certain types of quick calculations:

$$19 \cdot 11 = 19(10 + 1) = 19 \cdot 10 + 19 \cdot 1$$
$$= 190 + 19 = 209,$$
$$23 \cdot 9 = 23(10 - 1) = 230 - 23 = 207.$$

Also, if the first form of the distributive law is written in reverse order, as $ab + ac = a(b + c)$, then the two terms on the left have $a$ as a common factor, which is "factored out" on the right. The same comment applies to the other forms. This is an important technique for simplifying algebraic expressions:

$$16abc + 8ac - 280ad = 8a(2bc + c - 35d);$$
$$32(a-b)a - 142(a-b)^2 = 2(a-b)$$
$$\times [16a - 71(a-b)]$$
$$= 2(a-b)$$
$$\times [16a - 71a + 71b]$$
$$= 2(a-b)$$
$$\times [71b - 55a].$$

(5) Fractions are added and subtracted by the following rules:

$$\frac{a}{b} + \frac{c}{d} = \frac{ad + bc}{bd}$$ (cross-multiply and add to get the numerator),

$$\frac{a}{b} - \frac{c}{d} = \frac{ad - bc}{bd}$$ (cross-multiply and subtract to get the numerator).

The rules for multiplication and division are

$$\frac{a}{b} \cdot \frac{c}{d} = \frac{ac}{bd} \quad \text{and} \quad \frac{\frac{a}{b}}{\frac{c}{d}} = \frac{a}{b} \cdot \frac{d}{c}.$$

This division rule is often expressed in words as follows: to divide, invert the denominator and multiply.

**Examples.**

$$\frac{3a}{2} - \frac{4a}{3} = \frac{9a - 8a}{6} = \frac{a}{6};$$

$$\frac{1}{a+b} + \frac{1}{a-b} = \frac{(a-b) + (a+b)}{(a+b)(a-b)} = \frac{2a}{a^2 - b^2};$$

$$\frac{\frac{18abc}{2d}}{\frac{3}{ad}} = \frac{18abc}{2d} \cdot \frac{ad}{3} = \frac{18a^2bcd}{6d} = 3a^2bc.$$

# EXERCISES

2. Remove parentheses etc. and simplify:
   (a) $(3a - b) - [2a - (a+b)]$;
   (b) $[(a + 3b) - a] - [a - (a - 3b)]$;
   (c) $a - \{2a - [b - (3a - 2b)]\}$.
3. Multiply by using a suitable form of the distributive law: (a) $19 \cdot 179$; (b) $510 \cdot 18$; (c) $302 \cdot 11$.
4. Simplify by removing common factors:
   (a) $12x - 18y + 30$;

   (b) $8x^2 - 12x^3y - 28x^4z$;

   (c) $9abc + 3a^2b^2c^2$.

5.   Combine and simplify:

   (a) $\dfrac{a}{b} - \dfrac{b}{a}$;

   (b) $\dfrac{3}{x-2} + \dfrac{1}{2-x}$;

   (c) $\dfrac{1}{1 + \dfrac{1}{x-1}}$;

   (d) $\dfrac{x}{xy^2} + \dfrac{y}{x^2y}$;

   (e) $\dfrac{4a}{b} + \dfrac{b}{4a}$.

# 1.3 INTEGRAL EXPONENTS

We have already used the exponent notation for squares and cubes and fourth powers, $a^2 = a \cdot a$, $a^3 = a \cdot a \cdot a$, and $a^4 = a \cdot a \cdot a \cdot a$; and in the same way, $a^n$ is defined by $a^n = a \cdot a \cdots a$ ($n$ factors) for any positive integer $n$. The rules of exponents are quite simple:

| Rule | Illustration and explanation |
|------|------------------------------|

(1) $a^m a^n = a^{m+n}$ $\qquad$ $a^2 a^3 = \underbrace{(a \cdot a)}_{2}\underbrace{(a \cdot a \cdot a)}_{3} = \underbrace{a \cdot a \cdot a \cdot a \cdot a}_{5 \text{ factors}} = a^5$

(2) $\dfrac{a^m}{a^n} = a^{m-n}$ $\qquad$ $\dfrac{a^5}{a^3} = \dfrac{a \cdot a \cdot a \cdot a \cdot a}{a \cdot a \cdot a} = \dfrac{a \cdot a \cdot a}{a \cdot a \cdot a} \cdot \dfrac{a \cdot a}{1} = a \cdot a = a^2$

(3) $(a^m)^n = a^{mn}$ $\qquad$ $(a^3)^2 = (a \cdot a \cdot a)(a \cdot a \cdot a) = a \cdot a \cdot a \cdot a \cdot a \cdot a = a^6$

(4) $(ab)^n = a^n b^n$ $\qquad$ $(ab)^3 = (ab)(ab)(ab) = a \cdot a \cdot a \cdot b \cdot b \cdot b = a^3 b^3$

(5) $\left(\dfrac{a}{b}\right)^n = \dfrac{a^n}{b^n}$ $\qquad$ $\left(\dfrac{a}{b}\right)^4 = \left(\dfrac{a}{b}\right)\left(\dfrac{a}{b}\right)\left(\dfrac{a}{b}\right)\left(\dfrac{a}{b}\right) = \dfrac{a \cdot a \cdot a \cdot a}{b \cdot b \cdot b \cdot b} = \dfrac{a^4}{b^4}$

If $a \neq 0$, rule (1) suggests a natural way to define $a^0$. Since we want the equation $a^0 a^n = a^{0+n} = a^n$ to be true, $a^0$ should have the property that it leaves $a^n$ unchanged by multiplication. We therefore *define* $a^0$ by $a^0 = 1$. Similarly, we want the equation $a^n a^{-n} = a^{n-n} = a^0 = 1$ to be true, so $a^{-n}$ ought to be the reciprocal of $a^n$ and we *define* $a^{-n}$ by $a^{-n} = \dfrac{1}{a^n}$. With these definitions, the rules of exponents remain valid for all integral exponents, positive, negative and zero.

**Example.** Write each of the following with positive exponents:

$$a^6b^{-4}, \frac{1}{a^{-5}}, \frac{a^{-3}}{b^2}.$$

**Solution.** $a^6b^{-4} = a^6 \cdot \frac{1}{b^4} = \frac{a^6}{b^4}, \frac{1}{a^{-5}} = \frac{1}{\frac{1}{a^5}} = a^5,$

$$\frac{a^{-3}}{b^2} = \frac{\frac{1}{a^3}}{b^2} = \frac{1}{a^3b^2}.$$

As this example illustrates, any factor $a^n$ in a product or quotient can be moved from the numerator to the denominator or vice versa, if we change the sign of the exponent.

# EXERCISES

6. Simplify by removing negative and zero exponents: (a) $5a^{-3}$; (b) $(5a)^{-3}$; (c) $21 \cdot 719^3 \cdot 7^{-1} \cdot 3 \cdot 719^{-3}$;

   (d) $\left[\frac{2a^{-3} + 3a^{-2}}{3a^{-4} + 4a^{-3}}\right]^0$.

7. Simplify: (a) $(a^{n-4}b^4)(ab^{n-1})^4$; (b) $(4a^3b^{-4})$
   $\cdot (3a^{-1}b^5)$; (c) $\frac{x^{14}y^5}{x^4y^{-5}}$; (d) $a^2b^2(a^{-2} + b^{-2})$;
   (e) $(x + y)(x^{-1} + y^{-1})$;

   (f) $\left(\frac{a^2b}{c}\right)^4\left(\frac{a}{b^2c^3}\right)^2\left(\frac{c^2}{a^2}\right)^5$.

# 1.4 ROOTS AND RADICALS

Even though we have already used the notation of radicals in mentioning such numbers as $\sqrt{2}$ and $\sqrt{3}$, it will be useful to repeat and generalize the underlying ideas.

If $n$ is a positive integer and $x^n = a$, then $x$ is called an *nth root* of $a$. In particular, $x$ is called a *square root* of $a$ if $x^2 = a$, and a *cube root* of $a$ if $x^3 = a$.

**Example.** $2^2 = 4$ and $(-2)^2 = 4$, so 2 and $-2$ are both square roots of 4; $2^3 = 8$, so 2 is a cube root of 8; $(-2)^3 = -8$, so $-2$ is a cube root of $-8$; $3^4 = 81$ and $(-3)^4 = 81$, so 3 and $-3$ are both fourth roots of 81; $2^5 = 32$, so 2 is a fifth root of 32.

Since the square of a positive or negative real number is positive, negative numbers have no real square roots. However, each positive number $a$ has two square roots, numerically equal but opposite in sign, and the positive one of these is denoted by $\sqrt{a}$. Similarly, if $a$ is positive it has a single positive $n$th root denoted by $\sqrt[n]{a}$, and if $a$ is negative and $n$ is odd it has a single negative $n$th root denoted by $\sqrt[n]{a}$. The symbol $\sqrt{\phantom{a}}$ is called the *radical sign*, and the number $n$ placed above the radical sign is called the *index* of the root. As indicated above, in the case of a square root ($n = 2$) the index is omitted.

**Example.** $\sqrt{36} = 6$, $\sqrt[3]{-27} = -3$, $\sqrt[4]{16} = 2$, $\sqrt[5]{-1} = -1$, $\sqrt[6]{-1}$ does not exist. This example illustrates the following basic facts: *if a is positive and* n *is even or odd,* $\sqrt[n]{a}$ *is positive; if a is negative and* n *is odd,* $\sqrt[n]{a}$ *is negative; if a is negative and* n *is even,* $\sqrt[n]{a}$ *does not exist.*

The main rules for the manipulation of radicals can be stated as follows:

|  | Rule | Illustration |
|---|---|---|
| (1) | $(\sqrt[n]{a})^n = a$ | $(\sqrt{2})^2 = 2$ |
| (2) | $\sqrt[n]{a^n} = a$ | $\sqrt[3]{125} = \sqrt[3]{5^3} = 5$ |
| (3) | $\sqrt[n]{ab} = \sqrt[n]{a} \cdot \sqrt[n]{b}$ | $\sqrt{50} = \sqrt{25 \cdot 2} = \sqrt{25} \cdot \sqrt{2} = 5\sqrt{2}$ |
| (4) | $\sqrt[n]{\dfrac{a}{b}} = \dfrac{\sqrt[n]{a}}{\sqrt[n]{b}}$ | $\sqrt{\dfrac{4}{9}} = \dfrac{\sqrt{4}}{\sqrt{9}} = \dfrac{2}{3}$ |
| (5) | $\sqrt[m]{\sqrt[n]{a}} = \sqrt[mn]{a}$ | $\sqrt[6]{a^2} = \sqrt[3]{\sqrt{a^2}} = \sqrt[3]{a}$ |

In computational work it is often convenient to remove square roots from denominators of fractional expressions. The following calculations demonstrate two ways of doing this:

$$\frac{2}{\sqrt{2}} = \frac{2}{\sqrt{2}} \cdot \frac{\sqrt{2}}{\sqrt{2}} = \frac{2\sqrt{2}}{2} = \sqrt{2},$$

$$\frac{2}{\sqrt{3}+1} = \frac{2}{\sqrt{3}+1} \cdot \frac{\sqrt{3}-1}{\sqrt{3}-1} = \frac{2(\sqrt{3}-1)}{(\sqrt{3})^2 - 1^2}$$

$$= \frac{2(\sqrt{3}-1)}{3-1} = \sqrt{3} - 1.$$

These procedures are called *rationalizing the denominator.*

## EXERCISES

8. Simplify: (a) $\sqrt{49}$; (b) $\sqrt{144}$; (c) $\sqrt{9+16}$; (d) $\sqrt{36+64}$; (e) $\sqrt[3]{27}$; (f) $\sqrt[4]{81}$; (g) $\sqrt[6]{64}$;

(h) $\sqrt{.64}$; (i) $\sqrt{.09}$; (j) $\sqrt{\dfrac{16}{121}}$; (k) $\sqrt{\dfrac{225}{400}}$;

(l) $\sqrt[3]{-\dfrac{1}{27}}$; (m) $\sqrt[3]{\dfrac{64}{125}}$; (n) $\sqrt[3]{-1000}$; (o) $\sqrt{125}$;

(p) $\sqrt{625}$; (q) $\sqrt[4]{625}$; (r) $\sqrt{18}$; (s) $\sqrt{12}$; (t) $\sqrt{2}$
$+ \sqrt{8}$; (u) $\sqrt{3} + \sqrt[4]{9}$; (v) $\sqrt[3]{54} + \sqrt[3]{250}$;
(w) $\sqrt[10]{32a^5}$; (x) $\sqrt{a^2b^4}$; (y) $\sqrt[4]{a^5}$;

(z) $\sqrt{1 - \left(\dfrac{\sqrt{3}}{2}\right)^2}$.

9. Simplify by rationalizing the denominator:
   (a) $\dfrac{30}{\sqrt{6}}$; (b) $\dfrac{\sqrt{6}+2}{\sqrt{6}-2}$; (c) $\dfrac{2}{\sqrt{7}+\sqrt{5}}$.

# 1.5 FRACTIONAL EXPONENTS

Fractional exponents are a great convenience — almost a necessity — in many parts of mathematics. Our purpose here is to construct the definitions in such a way that the general rules of exponents given in Section 1.3 remain valid.

First, we want the equation $(a^{1/2})^2 = a^{(1/2) \cdot 2} = a^1 = a$ to be true, and this tells us that $a^{1/2}$ must be defined as follows: $a^{1/2} = \sqrt{a}$. Similarly, we define $a^{1/3}$ to be $\sqrt[3]{a}$, and $a^{1/n}$ to be $\sqrt[n]{a}$ for any positive integer $n$.

**Example.** $9^{1/2} = \sqrt{9} = 3$, $64^{1/3} = \sqrt[3]{64} = 4$, $(-27)^{1/3} = \sqrt[3]{-27} = -3$, $16^{1/4} = \sqrt[4]{16} = 2$.

We next assume that any fraction used as an exponent is written in its so-called lowest terms, that is, in the form $m/n$ where $n$ is a positive integer, $m$ is an integer (positive, negative or zero), and $m$ and $n$ have no common factor greater than 1. We want it to be true that $(a^{m/n})^n = a^m$, so that $a^{m/n}$ is an $n$th root of $a^m$. We therefore define $a^{m/n}$ to be $\sqrt[n]{a^m}$.

**Example.** $4^{3/2} = \sqrt{4^3} = \sqrt{64} = 8$, $8^{2/3} = \sqrt[3]{8^2} = \sqrt[3]{64} = 4$, $a^{2/7} \cdot a^{5/2} = a^{4/14} \cdot a^{35/14} = a^{39/14} = \sqrt[14]{a^{39}}$,
$\dfrac{a^{2/3}}{a^{3/5}} = a^{(2/3 - 3/5)} = a^{1/15} = \sqrt[15]{a}$, $\dfrac{\sqrt[3]{a^4}}{\sqrt[4]{a^3}} = \dfrac{a^{4/3}}{a^{3/4}} = a^{(4/3 - 3/4)}$
$= a^{7/12} = \sqrt[12]{a^7}$.

It is occasionally useful to know that $\sqrt[n]{a^m} = (\sqrt[n]{a})^m$, so we have
$a^{m/n} = \sqrt[n]{a^m} = (\sqrt[n]{a})^m$.

This is not difficult to prove, but we omit the details.

**Example.** $8^{2/3}$ is easy to evaluate both ways, for (as we noted above) $8^{2/3} = \sqrt[3]{8^2} = \sqrt[3]{64} = 4$ and also $8^{2/3} = (\sqrt[3]{8})^2 = 2^2 = 4$. However, $32^{3/5} = \sqrt[5]{32^3}$ is hard to evaluate but $32^{3/5} = (\sqrt[5]{32})^3 = 2^3 = 8$ is easy.

## EXERCISES

10. Compute: (a) $36^{1/2}$; (b) $8^{1/3}$; (c) $32^{4/5}$; (d) $36^{3/2}$; (e) $216^{2/3}$; (f) $16^{-1/2}$; (g) $9^{-3/2}$; (h) $8^{-2/3}$; (i) $100^{3/2}$; (j) $3^{1/2} \cdot 3^{5/2}$; (k) $\dfrac{10^{2/3} \cdot 10^{1/3} \cdot 10^3}{10^{5/2} \cdot 10^{1/2}}$.

11. Simplify as far as possible, removing negative and zero exponents: (a) $(25a^6b^{-2})^{1/2}$; (b) $(2a^{1/2}b^{1/4})^4$; (c) $\sqrt[5]{a^2b} \cdot \sqrt[5]{a^3b^4}$; (d) $\left(\dfrac{a^4}{36}\right)^{1/2}$; (e) $(25a^{2/3})^{1/2}$; (f) $(a^{1/2} + b^{1/2})(a^{1/2} - b^{1/2})$; (g) $\left\{a^{2/3}\left[\left(\dfrac{a^{2/3}}{a^{1/4}}\right)^6\right]^{1/3}\right\}^2$; (h) $\left(\dfrac{27b^2c^5}{64a^6b^{-4}c^{-1}}\right)^{1/3}$.

## 1.6 POLYNOMIALS

A letter which is permitted to represent different numbers during a single discussion is called a *variable*. A symbol which represents only one specific number throughout a discussion is called a *constant*. As we stated in Section 1.2, it is customary to use the later letters of the alphabet for variables and the earlier ones for constants. Thus, the letters $x$, $y$, $z$ are commonly used for variables and $a$, $b$, $c$ for constants. Of course, specific numbers like 7, $\sqrt{2}$ and $\pi$ are also constants.

A *polynomial*, or more precisely, a *polynomial in x*, is an algebraic expression that can be built up from the variable $x$ and any constants by means of the operations of addition, subtraction and multiplication alone. For example,

$$1, \quad x, \quad x + 1, \quad 2x^2, \quad 2x^2 + x + 1$$

and

$$3x^5 - 7x^4 + 2x^3 + 13x^2 - 4x + 22$$

are polynomials. In the last example, the constants 3, $-7$, 2, 13, $-4$ and 22 are called the *coefficients* of the polynomial, and the exponent of the highest power of $x$—in this case, 5—is called its

**43**

*degree*. Certain polynomials have special names according to their degrees:

*constant polynomial* (degree 0): $a$ ($a \neq 0$).

*linear polynomial* (degree 1): $ax + b$ ($a \neq 0$).

*quadratic polynomial* (degree 2): $ax^2 + bx + c$ ($a \neq 0$).

*cubic polynomial* (degree 3): $ax^3 + bx^2 + cx + d$ ($a \neq 0$).

The constant polynomial 0 is not assigned a degree.

Polynomials are added and subtracted by the simple device of using inspection to combine terms involving the same powers of $x$.

**Examples.**

$(3x^3 - 7x^2 + 13x - 2) + (9x^2 + 4x + 19)$
$= 3x^3 + 2x^2 + 17x + 17;$
$(5x^4 + 2x^2 - 3) - (6x^3 - 2x^2 + 4x + 6)$
$= 5x^4 - 6x^3 + 4x^2 - 4x - 9.$

Polynomials are multiplied like any other sums. If there are two polynomial factors, we multiply the second factor by each term of the first, simplify by using the exponent rule $x^m \cdot x^n = x^{m+n}$, and collect terms involving the same powers of $x$.

**Example.**

$(2x^2 - 4x + 3)(3x^3 - x^2 + 8x - 5)$
$= 6x^5 - \phantom{0}2x^4 + 16x^3 - 10x^2$
$\phantom{= 6x^5} - 12x^4 + \phantom{0}4x^3 - 32x^2 + 20x$
$\phantom{= 6x^5 - 12x^4} + \phantom{0}9x^3 - \phantom{0}3x^2 + 24x - 15$
$= 6x^5 - 14x^4 + 29x^3 - 45x^2 + 44x - 15.$

It is clear that the degree of the product of two nonzero polynomials equals the sum of their individual degrees.

# EXERCISES

12. Add or subtract:
    (a) $(x^7 - 3x^5 + 4x^2 - 9)$
        $+ (2x^6 - 5x^5 - 2x^4 + x^3 - 2x^2 + x + 1);$
    (b) $(3x^5 + x^4 - 2x^3 + 5x^2 - 11x + 2)$
        $- (x^4 + 5x^2 + 2).$

13. Multiply:
    (a) $(2x^3 + 3x^2 - 4)(3x^2 - 2x - 9);$
    (b) $(x^5 - 2x^3 + 3)(2x^2 - 8x + 4);$

    (c) $(x-1)(x^2+x+1)$;

    (d) $(x-1)(x^3+x^2+x+1)$;

    (e) $(x-1)(x^4+x^3+x^2+x+1)$.

## 1.7 FACTORING

To factor a polynomial is to express it as a product of polynomials of lower degrees. We will see in the next section that factoring is a useful method for solving certain kinds of equations.

The simplest type of factoring is removing a common polynomial factor, and this should always be done first.

**Examples.**

$x^4+x^2 = x^2(x^2+1)$;

$3x^5-39x^4+15x^3 = 3x^3(x^2-13x+5)$;

$$2(x-5)^3-6(x-5)^2 = 2(x-5)^2[(x-5)-3]$$
$$= 2(x-5)^2(x-8).$$

Most factoring depends on recognizing the expanded forms of certain special products:

(1)   $(x+a)(x-a) = x^2-a^2$;

(2)   $(x+a)(x+a) = (x+a)^2 = x^2+2ax+a^2$;

(3)   $(x-a)(x-a) = (x-a)^2 = x^2-2ax+a^2$;

(4)   $(x+a)(x+b) = x^2+(a+b)x+ab$;

(5)   $(ax+b)(cx+d) = acx^2+(ad+bc)x+bd$.

These products occur so often in algebraic computation that they should be verified by the student and memorized. Reading from left to right, each can be considered as an expansion formula; but when they are read from right to left they are *factoring formulas*, and this is their real importance.

**Examples.** The following illustrate (1), (2) and (3):

$x^2-25 = (x+5)(x-5)$;

$4x^2-9 = (2x)^2-3^2 = (2x+3)(2x-3)$;

$x^2+6x+9 = (x+3)^2$;

$x^2-10x+25 = (x-5)^2$.

In using (4) to factor a polynomial of the form $x^2+px+q$, the trick is to think of various pairs of numbers $a$ and $b$ whose product is $q$, and hope to find one such pair whose sum is $p$.

**Examples.**

$x^2 + x - 6 = (x + 3)(x - 2);$
$x^2 + 10x + 21 = (x + 3)(x + 7);$
$x^2 - 9x + 18 = (x - 3)(x - 6).$

The uses of (5) are similar but more difficult, because in most cases several possibilities must be tested before the right combination is found.

**Examples.**

$2x^2 + 5x - 3 = (2x - 1)(x + 3);$
$3x^2 + x - 2 = (x + 1)(3x - 2);$
$12x^2 + 7x - 10 = (3x - 2)(4x + 5);$
$8x^2 + 10x - 12 = (4x - 3)(2x + 4).$

The successful application of this factoring method is clearly a matter of trial and error, and a good supply of patience is necessary.

# EXERCISES

14. Factor: (a) $x^2 - x - 6$; (b) $x^2 + 9x + 20$; (c) $x^2 + 12x + 20$; (d) $x^2 - 4x + 4$; (e) $x^2 + 8x + 16$; (f) $x^3 + 12x^2 + 36x$; (g) $x^4 - 16$; (h) $x^2 + 13x - 30$; (i) $x^2 + 2x - 35$; (j) $x^2 - 13x + 42$; (k) $x^3 - 3x^2 - 4x$; (l) $4x^2 + 2x - 12$; (m) $10x^2 - 16x - 8$.

15. Verify the formula $(x - a)(x^2 + ax + a^2) = x^3 - a^3$ and use it to factor (a) $x^3 - 27$; (b) $8x^3 - 125$.

16. Verify the formula $(x + a)(x^2 - ax + a^2) = x^3 + a^3$ and use it to factor (a) $x^3 + 64$; (b) $27x^3 + 8$.

# 1.8 LINEAR AND QUADRATIC EQUATIONS

*Linear equations* are equations like

$3x - 12 = 0$ and $2x + 18 = 0,$

and the general form is

$ax + b = 0, \qquad a \neq 0.$

The method of solution is very easy: move the constant term $b$ to the right,

$ax = -b,$

and then move the coefficient $a$ into the denominator on the right,

$x = -\dfrac{b}{a}.$

In effect, we isolate the unknown $x$ (solve for $x$) by first subtracting $b$ from both sides and then dividing both sides by $a$. The solutions of the two equations at the beginning of the paragraph are easily seen by inspection to be $x = 4$ and $x = -9$.

*Quadratic equations* are equations like

$$x^2 + x - 12 = 0, \tag{1}$$

and the general form is

$$ax^2 + bx + c = 0, \qquad a \neq 0. \tag{2}$$

Such equations can sometimes be solved by factoring. Thus, the factored form of equation (1) is

$$(x + 4)(x - 3) = 0, \tag{3}$$

and we obtain its roots (solutions) by setting each factor equal to zero:

$$x + 4 = 0, \text{ so } x = -4; \qquad x - 3 = 0, \text{ so } x = 3.$$

This procedure for finding the roots $-4$ and $3$ rests on the following basic principle: a product of two numbers equals zero if and only if one of the factors is zero. This tells us that equation (1) — or equivalently (3) — is satisfied if and only if $x + 4 = 0$ or $x - 3 = 0$, and each of these equations yields one of our roots.

If factoring fails, we can always fall back on the following general formula — called the *quadratic formula* — for the roots of (2) in all possible cases:

$$x = \frac{-b \pm \sqrt{b^2 - 4ac}}{2a}. \tag{4}$$

If this is applied directly to (1), with $a = 1$, $b = 1$ and $c = -12$, we get

$$x = \frac{-1 \pm \sqrt{1 + 48}}{2} = \frac{-1 \pm \sqrt{49}}{2}$$

$$= \frac{-1 \pm 7}{2} = 3 \text{ or } -4,$$

as before.

In order to establish (4) we begin by examining the following formula for the square of the quantity $x + A$:

$$(x + A)^2 = x^2 + 2Ax + A^2.$$

The right side of this is evidently a perfect square — the square of $x + A$ — because *its constant term is the square of half the coefficient of* $x$. This remark is the basis for an algebraic method called *completing the square*, which is useful in many parts of mathematics.

We now derive the quadratic formula (4) by applying the following operations to equation (2): divide by $a$,

$$x^2 + \frac{b}{a}x + \frac{c}{a} = 0;$$

move the constant term to the right,

$$x^2 + \frac{b}{a}x = -\frac{c}{a};$$

complete the square on the left by adding $\left(\frac{b}{2a}\right)^2$

$= \frac{b^2}{4a^2}$ [the square of half the coefficient of $x$], and add the same number to the right, obtaining

$$x^2 + \frac{b}{a}x + \frac{b^2}{4a^2} = -\frac{c}{a} + \frac{b^2}{4a^2},$$

or equivalently

$$\left(x + \frac{b}{2a}\right)^2 = \frac{b^2 - 4ac}{4a^2};$$

and finally, take square roots,

$$x + \frac{b}{2a} = \frac{\pm\sqrt{b^2 - 4ac}}{2a},$$

and solve for $x$—and the result is (4).

**Examples.** If we apply (4) to $x^2 - 6x + 6 = 0$, then $a = 1$, $b = -6$ and $c = 6$, and the roots are

$$x = \frac{6 \pm \sqrt{36 - 24}}{2} = \frac{6 \pm 2\sqrt{3}}{2} = 3 \pm \sqrt{3}.$$

If we apply (4) to $x^2 + 4x + 4 = 0$, then $a = 1$, $b = 4$ and $c = 4$, and the roots are

$$x = \frac{-4 \pm \sqrt{16 - 16}}{2} = \frac{-4 \pm 0}{2} = -2, -2.$$

If we apply (4) to $x^2 - 2x + 2 = 0$, then $a = 1$, $b = -2$ and $c = 2$, and the roots are

$$x = \frac{2 \pm \sqrt{4 - 8}}{2} = \frac{2 \pm 2\sqrt{-1}}{2} = 1 \pm \sqrt{-1}.$$

The three parts of this example display the three possibilities for the roots of a quadratic equation: two distinct real roots, two equal real roots, and two distinct imaginary roots. The third possibility involves square roots of negative numbers, which are not real numbers and are beyond the scope of this discussion. The geometric meaning of these possibilities will be made clear in Section 2.6 below, in connection with the graphs of quadratic functions.

## EXERCISES

17. Solve by factoring, and then by the quadratic formula: (a) $x^2 + 3x - 28 = 0$; (b) $x^2 - 8x - 33 = 0$; (c) $2x^2 + x - 15 = 0$; (d) $6x^2 - 5x - 21 = 0$.

18. Solve by the quadratic formula: (a) $5x^2 - 9x + 3 = 0$; (b) $3x^2 + 7x + 3 = 0$; (c) $17x^2 - 6x + 1 = 0$; (d) $x^2 + x + 1 = 0$.

## 1.9 INEQUALITIES AND ABSOLUTE VALUES

The left-to-right linear arrangement of points on the real line (see Fig. 2) corresponds to the part of algebra dealing with inequalities. If we visualize Fig. 2 and think of real numbers as points on the real line, then the essential ideas can be stated as follows.

The inequality $a < b$ (read "$a$ is less than $b$") means that the point $a$ lies to the left of the point $b$, and the equivalent inequality $b > a$ ("$b$ is greater than $a$") means that $b$ lies to the right of $a$. A number $a$ is positive or negative according as $a > 0$ or $a < 0$. The main rules used in working with inequalities are these:

(1) if $a > 0$ and $b < c$, then $ab < ac$;

(2) if $a < 0$ and $b < c$, then $ab > ac$;

(3) if $a < b$, then $a + c < b + c$ for any $c$.

Rules (1) and (2) are usually expressed by saying that an inequality is preserved on multiplication by a positive number, and reversed on multiplication by a negative number; and (3) says that an inequality is preserved when any number (positive or negative) is added to both sides.

The statement that $a$ is positive or equal to zero is written $a \geq 0$, and is read "$a$ is greater than or equal to zero." In the same way, $a \geq b$ means $a > b$ or $a = b$. Thus, $5 \geq 3$ and $5 \geq 5$ are both true inequalities.

It is often useful to know that a product of nonzero numbers is positive if it has an even number of negative factors, and negative if it has an odd number of negative factors.

**Example.** To "solve" a linear inequality like
$$3x - 2 < 6 - x$$
means to find the values of the variable $x$ for which the inequality is true. By using the above rules, the following inequalities are seen to be equivalent to the one given:

$4x - 2 < 6$      (add $x$ to both sides);

$4x < 8$      (add 2 to both sides);

$x < 2$      $\left(\text{multiply both sides by } \frac{1}{4}\right).$

The last inequality is the solution.

The *absolute value* of a number $a$ is denoted by the symbol $|a|$ (read "absolute value of $a$") and is defined by
$$|a| = \begin{cases} a \text{ if } a \geq 0, \\ -a \text{ if } a < 0. \end{cases}$$
For example, $|3| = 3$, $|-5| = -(-5) = 5$, and $|0| = 0$. It is easy to see that the operation of forming the absolute value leaves positive numbers unchanged, and replaces each negative number by the corresponding positive number. The main properties of this operation are
$$|a| \geq 0, \quad |ab| = |a||b|, \quad |a + b| \leq |a| + |b|.$$
In the language of Fig. 2, the absolute value of a number $a$ is simply the distance from the point $a$ to the origin. In the same way, the distance from $a$ to $b$ is $|a - b|$.

# EXERCISES

19. Insert the correct inequality sign, $<$ or $>$, between each of the following pairs of numbers: (a) 3 ___ 11; (b) 5 ___ 2; (c) $-4$ ___ 3; (d) $-6$ ___ $-2$; (e) $-2$ ___ $-3$; (f) $\pi$ ___ $2\sqrt{3}$.

20. Solve the linear inequalities (a) $5 - 2x > 17$; (b) $3x + 4 > 13$.

21. Solve the equations (a) $|x| = 2$; (b) $|2x| = 6$; (c) $\left|\frac{1}{3}x\right| = 2$; (d) $|x - 2| = 3$; (e) $|x + 3| = 1$.

22. Solve the quadratic inequality $x(x - 1) > 0$ by noticing that both factors must be positive or both factors must be negative.

23. Solve the inequality $x^2 + 2x - 15 > 0$ by using the idea suggested in the preceding exercise.

# 2. FUNCTIONS AND GRAPHS
## 2.1 THE CONCEPT OF A FUNCTION

Everyone knows about the use of graphs to summarize data and convey ideas (Fig. 3). The

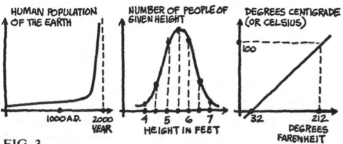

FIG. 3

figure shows three graphs whose meanings are reasonably clear on inspection. There are many kinds of graphs, dealing with many different subjects, but all have one thing in common: each is a visual display showing the way in which one variable quantity depends on another. Graphs are pictures of functions, and functions are the basic ingredients of quantitative knowledge.

What is a function? If $x$ and $y$ are two variables that are related in such a way that whenever a permissible numerical value is assigned to $x$ there is determined one and only one corresponding numerical value for $y$, then $y$ is called a *function of x.*

**Examples.** (a) If a rock is dropped from the edge of a cliff, then it falls $s$ feet in $t$ seconds, and $s$ is a function of $t$. It has been found by experiment that (approximately) $s = 16t^2$.

(b) The area $A$ of a circle is a function of its radius $r$, and it is known from geometry that $A = \pi r^2$.

(c) If the manager of a bookstore buys $n$ books from a publisher at \$8 each and the shipping charges are \$13, then his total cost $C$ is a function of $n$ given by the formula $C = 8n + 13$.

If $y$ is a function of $x$ as explained above, it is customary to express this in symbols by writing

$$y = f(x).$$

This is read "$y$ equals $f$ of $x$," and the letter $f$ represents the formula or rule of correspondence

51

that yields $y$ when applied to $x$. The variable $x$ is called the *independent variable*, and the variable $y$, whose value depends on the value of $x$, is called the *dependent variable*. If $a$ is a specific permissible value of $x$, then the number $f(a)$ is called the *value* of $f(x)$ at $x = a$.

There is nothing illegal or immoral about using other letters than $x$ and $y$ to denote the variables. In the above examples, for instance, the independent variables are $t$, $r$ and $n$, and the dependent variables are $s$, $A$ and $C$. Also, as we see in the next examples, other letters than $f$ can be used to designate functions.

**Examples.** (a) If a function $f(x)$ is defined by the formula $f(x) = x^3 - 3x + 2$, then $f(2) = 2^3 - 3 \cdot 2 + 2 = 4$, $f(0) = 2$ and $f(-1) = (-1)^3 - 3(-1) + 2 = 4$.

(b) If a function $g(x)$ is defined by the formula $g(x) = \sqrt{x}$, then $g(0) = \sqrt{0} = 0$, $g(4) = \sqrt{4} = 2$ and $g(9) = \sqrt{9} = 3$. In this case the only permissible values of $x$ are those for which $x \geq 0$.

(c) If a function $h(x)$ is defined by the formula $h(x) = \dfrac{1}{3-x}$, then $h(1) = \dfrac{1}{3-1} = \dfrac{1}{2}$, $h(2) = \dfrac{1}{3-2}$ $= 1$ and $h(3)$ does not exist—because division by zero is not allowed. Thus, $x = 3$ is the only value of $x$ that is not permissible.

Functions are often defined by formulas, as in the above examples. However, the most convenient way of visualizing a function is by means of its *graph*, as shown in Fig. 4. The independent variable $x$ can be thought of as a point moving along the $x$-axis from left to right; each $x$ determines a value of the dependent variable $y$, which is the height of the point $(x, y)$ above the $x$-axis; and the graph of the function is simply the path of the point $(x, y)$ as it moves across the plane and varies in height according to the nature of the particular function under discussion. We examine graphs in more detail in the next few sections.*

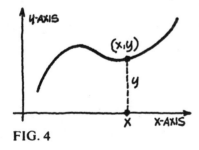

FIG. 4

---

*These remarks assume that the student already knows a bit about coordinates in a plane. If he does not, explanations are provided in the next section. In working with graphs in general, we are free to choose the units of measurement on

# EXERCISES

1. If $f(x) = 4x - 3$, find $f(0), f(1), f(2)$ and $f(3)$.

2. If $g(x) = \dfrac{2x - 4}{3x^2 + 1}$, find $g(0), g(1)$ and $g\left(-\dfrac{1}{2}\right)$.

3. If $h(x) = x^3 - 3x^2 + 5x - 1$, find $h(x^3)$.

4. If $F(x) = \dfrac{x}{x - 1}$, find $F[F(x)]$.

5. Express the area $A$ of a square as a function of (a) the length of one side $x$; (b) the perimeter $p$.

6. Express the area $A$ of a circle as a function of its circumference $c$.

7. Express the height of an equilateral triangle as a function of the base $x$.

8. Two bicyclists start racing along a straight road from the same place at the same time in the same direction. If one travels 40 mi/hr and the other 35 mi/hr, find the distance between them $t$ hours after they start.

## 2.2 COORDINATES IN A PLANE

We now leave the study of functions temporarily, and immerse ourselves for the next few sections in analytic geometry. The purpose of this detour is to gain some geometric background that will clarify the nature of some of the simpler functions and their graphs.

In Section 1.1 we used the real number system to assign coordinates to points on a line. Here we extend this procedure and assign coordinates to points in a plane, as shown in Fig. 5. A point called the *origin* is chosen in the plane, and two perpendicular lines are drawn through it, one horizontal and the other vertical. These lines are called the *x-axis* and *y-axis* respectively. We introduce a coordinate system on each axis in such a way that the unit distance is the same on both. Each point $P$ in the plane is assigned two coordinates as follows. Perpendicular lines are drawn from $P$ to the two axes. If these lines intersect the x-axis and y-axis at points having coordi-

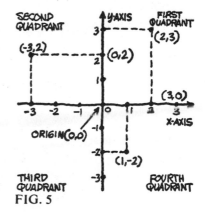

FIG. 5

each axis to suit our own convenience. However, in the following sections we use equal units on both axes because we wish to work with distances between points in the plane.

nates $x$ and $y$, then $x$ is called the *x-coordinate* and $y$ is called the *y-coordinate* of $P$. (Observe that if $P$ lies on the $x$-axis then $y = 0$, and if $P$ lies on the $y$-axis then $x = 0$.) In this way a one-to-one correspondence is established between all points $P$ in the plane and all ordered pairs $(x, y)$ of real numbers, and we consider points and ordered pairs to be essentially identical.

This system of assigning coordinates to points in a plane enables us to study geometry with the aid of the tools of algebra, and this is what analytic geometry is all about. We emphasize particularly that the same unit of length is used on both axes. This makes it possible to express the distance between two points in terms of their coordinates. Let $P_1 = (x_1, y_1)$ and $P_2 = (x_2, y_2)$ be the points, as shown in Fig. 6. Then the Pythagorean theorem applied to the indicated right triangle shows that

the distance between $P_1$ and $P_2$
$$= \sqrt{|x_1 - x_2|^2 + |y_1 - y_2|^2}$$
$$= \sqrt{(x_1 - x_2)^2 + (y_1 - y_2)^2}.$$

This is the *distance formula*, which is one of the most valuable tools of mathematics. In words, it says: *the distance between two points equals the square root of the sum of the squares of the differences of their coordinates.*

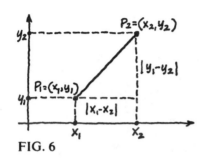

FIG. 6

# EXERCISES

9. Draw the triangle having the following points as vertices: (a) $(1, -1)$, $(4, 2)$, $(-3, 5)$; (b) $(5, -2)$, $(-2, -4)$, $(1, 4)$.

10. Find the length of the sides and the hypotenuse of the right triangle whose vertices are (a) $(1, 2)$, $(-3, 2)$, $(-3, 5)$; (b) $(2, -5)$, $(2, 7)$, $(-3, 7)$.

11. Find the area of the rectangle whose vertices are (a) $(4, 1)$, $(-2, 3)$, $(-2, 1)$, $(4, 3)$; (b) $(-3, 7)$, $(4, 2)$, $(-3, 2)$, $(4, 7)$.

12. Find the coordinates of the midpoint of the segment joining (a) $(0, 0)$ to $(6, 8)$; (b) $(1, 2)$ to $(7, 8)$.

13. Sketch on a suitable diagram all points $(x, y)$ such that (a) $x = 3$; (b) $y = -4$; (c) $x < 3$ and

$y > 2$; (d) $x$ or $y$ (or both) is zero; (e) $x \geq 0$ and $y \leq 0$; (f) $x^2 \leq 1$.

## 2.3 STRAIGHT LINES

If a curve is thought of as the path of a moving point, then straight lines are the simplest curves studied in analytic geometry, and they have the simplest equations.

This topic begins with the concept of the *slope* of a nonvertical line (Fig. 7). If $P_1 = (x_1, y_1)$ and $P_2 = (x_2, y_2)$ are any two distinct points on this line, then the slope of the line is denoted by $m$ and defined to be the ratio

$$m = \frac{y_2 - y_1}{x_2 - x_1}.$$

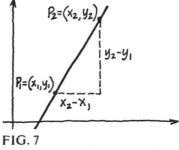

FIG. 7

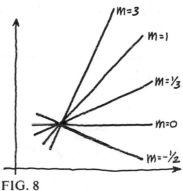

FIG. 8

The value of $m$ depends only on the line itself, and not on the particular positions of $P_1$ and $P_2$, and in the figure $m$ is the ratio of the height of the indicated triangle to its base. If $P_2$ is taken to be a point on the line one unit to the right of $P_1$, so that $x_2 - x_1 = 1$, then $m = \frac{y_2 - y_1}{1} = y_2 - y_1$. This shows that if we move one unit directly to the right from any point on the line, the slope $m$ can be thought of as the distance up or down we must move in order to get back on the line. Thus, the sign of the slope is related as follows to the direction of the line:

$m > 0$,    line rises to the right;

$m < 0$,    line falls to the right;

$m = 0$,    line horizontal.

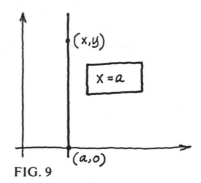

FIG. 9

Also, the numerical magnitude (absolute value) of the slope is a measure of the steepness of the line (Fig. 8).

We next discuss equations of straight lines. A vertical line is characterized by the fact that all points on it have the same $x$-coordinate, as shown in Fig. 9, and the $y$-coordinate is irrelevant. In Fig. 10 we consider a nonvertical line that is "given" in the sense that we know a point $(x_0, y_0)$ on it, and also its slope $m$. The boxed equation is the equation of the line for the reason suggested in the figure: it states an algebraic condition that is satisfied by the indicated variable point $(x, y)$

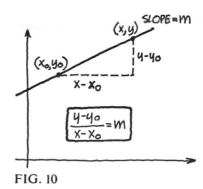

FIG. 10

on the line. These are two of the three primary forms for the equation of a straight line:

$$x = a, \qquad \frac{y - y_0}{x - x_0} = m, \qquad y = mx + b.$$

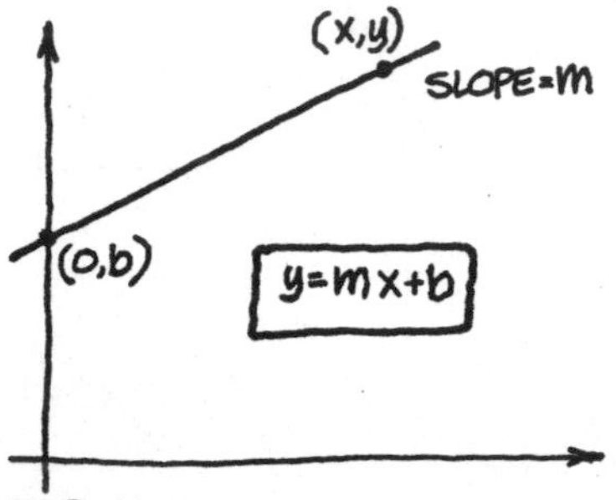

The third is simply the special case of the second that we get when the given point is a point $(0, b)$ on the $y$-axis, as shown in Fig. 11. In this case the number $b$ is called the *y-intercept* of the line, and the equation is called the *slope-intercept equation*. (The second equation is called the *point-slope equation*.)

Two lines with slopes $m_1$ and $m_2$ are clearly parallel if and only if their slopes are equal: $m_1 = m_2$. The criterion for perpendicularity is the equation

$$m_1 m_2 = -1.$$

This is not clear at all, but can be understood quite easily by using similar triangles, as follows (Fig. 12). Assume the lines are perpendicular, as shown in the figure. Draw a segment of length 1 to the right from their point of intersection, and from its right endpoint draw vertical segments up and down to the two lines. The meaning of the slopes tells us that the two right triangles formed in this way have sides of the indicated lengths. Since the lines are perpendicular. the indicated angles are equal, and therefore the triangles are similar and the following ratios of corresponding sides are equal:

$$\frac{m_1}{1} = \frac{1}{-m_2}, \text{ which is equivalent to } m_1 m_2 = -1.$$

The reasoning given here is easily reversible, so this condition also implies perpendicularity.

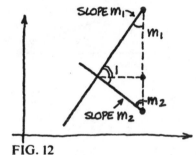

FIG. 11

FIG. 12

## EXERCISES

14. Find the slope of the line determined by (a) $(1, 2)$ and $(3, 6)$; (b) $(-2, 4)$ and $(3, 4)$; (c) $(-2, 1)$ and $(2, -3)$.

15. Find the equation of the line through (a) $(-1, -2)$ and $(1, 0)$; (b) $(0, 4)$ and $(1, 7)$; (c) $(4, -2)$ and $(4, 19)$.

16. Show that the lines $3x + y = 2$ and $2y = 1 - 6x$ are parallel.

17. Write the equation of the line through $(-2, -3)$ which is (a) parallel to $x + 2y = 3$; (b) perpendicular to $x + 2y = 3$.

## 2.4 CIRCLES

The *circle* with center $C$ and radius $r$ is defined to be the set of all points whose distance from $C$ is $r$. If the coordinates of $C$ are $(h, k)$ and the coordinates of a variable point on the circle are $(x, y)$, then by the distance formula of Section 2.2 the stated condition is

$$\sqrt{(x - h)^2 + (y - k)^2} = r,$$

and squaring to remove the radical yields the standard equation of the circle (Fig. 13):

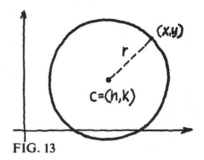
FIG. 13

$$(x - h)^2 + (y - k)^2 = r^2.$$

In particular,

$$x^2 + y^2 = r^2$$

is the equation of the circle with radius $r$ and center at the origin.

**Examples.** (a) The circle with center $(-3, 2)$ and radius 5 has $(x + 3)^2 + (y - 2)^2 = 25$ as its equation.

(b) The equation $(x - 4)^2 + (y + 2)^2 = 12$ is immediately recognizable as the equation of the circle with center $(4, -2)$ and radius $\sqrt{12} = 2\sqrt{3}$.

If the equation $(x + 3)^2 + (y - 2)^2 = 25$ is "simplified" by squaring and collecting terms, the result is the less informative — and therefore less useful — equation

$$x^2 + y^2 + 6x - 4y - 12 = 0.$$

In order to return to the standard form, and the information conveniently displayed in this form, we would have to complete the square on the variables $x$ and $y$.*

**Example.** To discover the nature of the equation

$$x^2 + y^2 - 10x + 6y + 18 = 0,$$

we rearrange it to prepare the way for completing the square:

$$(x^2 - 10x + \quad) + (y^2 + 6y + \quad) = -18.$$

---

*Recall that the structure of the equation $(x + a)^2 = x^2 + 2ax + a^2$ is the key to the process of completing the square; for the right side is a perfect square — the square of $x + a$ — precisely because its constant term is the square of half the coefficient of $x$.

The numbers that must be added inside the parentheses to make these quantities perfect squares are 25 and 9, so our equation becomes

$(x^2 - 10x + 25) + (y^2 + 6y + 9) = -18 + 25 + 9,$

or equivalently,

$(x - 5)^2 + (y + 3)^2 = 16.$

This is the equation of the circle with center $(5, -3)$ and radius 4.

It should be noted that if the constant term 18 in the first equation in this example is replaced by 34, then the final equation has 0 on the right side and is therefore the equation of the single point $(5, -3)$. This point can be thought of as the circle with center $(5, -3)$ and radius $r = 0$.

## EXERCISES

18. Write the equation of the circle (a) with center $(0, 0)$ and radius 2; (b) with center $(-2, 0)$ and radius 7; (c) with center $(3, 6)$ and radius $\frac{1}{2}$; (d) with $(5, 5)$ and $(-3, -1)$ as the ends of a diameter.

19. Find the center and radius of the circle whose equation is (a) $(x + 3)^2 + (y - 6)^2 = 9$; (b) $(x - 4)^2 + y^2 = 4$; (c) $x^2 + (y + 2)^2 = 1$; (d) $x^2 + y^2 + 6x + 2y + 6 = 0$; (e) $x^2 + y^2 - 16x + 14y + 97 = 0$.

20. On a single set of coordinate axes, sketch the line $x + 16 = 7y$ and the circle $x^2 + y^2 - 4x + 2y = 20$ and find their points of intersection. Hint: eliminate $x$ algebraically and solve the resulting quadratic equation in $y$.

## 2.5 PARABOLAS

A *parabola* is a plane curve with the property that each of its points is equidistant from a given fixed point and a given fixed line. The fixed point is called the *focus* and the fixed line is called the *directrix*. To find the simplest equation for this curve, let the focus be the point $F = (0, p)$ where $p$ is a positive number, and let the directrix be the line $y = -p$, as shown in Fig. 14. If $P = (x, y)$ is any point on the parabola, then by the definition we have $PF = PD$. Using the distance formula, this condition becomes

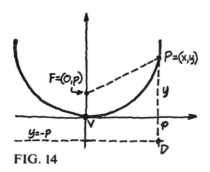

FIG. 14

58

$$\sqrt{x^2 + (y-p)^2} = y + p.$$

By squaring and simplifying we get

$$x^2 = 4py \quad \text{or} \quad y = \frac{1}{4p}x^2$$

as the equation of this parabola. The line through the focus perpendicular to the directrix is called the *axis* of the parabola, and the point $V$ where the parabola intersects the axis is called the *vertex*.

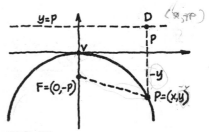

FIG. 15

The parabola in Fig. 14 opens up. However, if the focus and directrix are taken to be $F = (0, -p)$ and $y = p$, with $p$ still a positive number, as illustrated in Fig. 15, then the parabola opens down and a similar calculation shows that its equation is now

$$x^2 = -4py \quad \text{or} \quad y = -\frac{1}{4p}x^2.$$

For both parabolas the positive number $p$ is the distance from the vertex to the focus. In each case the parabola is symmetric about its axis, as we see geometrically and also from the fact that each equation is unchanged when $x$ is replaced by $-x$. In addition, we notice that the vertex is the low point of the parabola when it opens up, and the high point when it opens down.

**Example.** The equation $y = x^2$ is of the first type with $4p = 1$, so $p = \frac{1}{4}$. It therefore represents the parabola with vertex at the origin which opens up and has focus $\left(0, \frac{1}{4}\right)$ and directrix $y = -\frac{1}{4}$.

We illustrate a further point about parabolas by considering the equation

$$y = x^2 - 6x + 11.$$

If this is written in the form

$$y - 11 = x^2 - 6x,$$

and if we complete the square on the terms involving $x$, then the result is

$$y - 2 = (x - 3)^2.$$

If we now introduce new variables $\bar{x}$ and $\bar{y}$ by writing

$$\bar{x} = x - 3,$$
$$\bar{y} = y - 2,$$

then our equation becomes

$$\bar{y} = \bar{x}^2.$$

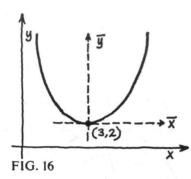

FIG. 16

The graph of this equation is clearly a parabola with vertex at the origin of the $\bar{x}$, $\bar{y}$ coordinate system, and this origin is located at the point $(3, 2)$ in the $x$, $y$ system, as shown in Fig. 16. In exactly the same way, we see that any equation of the form

$$y = ax^2 + bx + c \qquad (a \neq 0)$$

represents a parabola with vertical axis which opens up if $a > 0$ and down if $a < 0$. The vertex of this parabola is easily located by completing the square on $x$, and the equation can be written in the form

$$(x - h)^2 = 4p(y - k) \text{ or } (x - h)^2 = -4p(y - k),$$

where the point $(h, k)$ is the vertex.

## EXERCISES

21. Find the focus and directrix of each of the following parabolas: (a) $y = 2x^2$; (b) $y = \frac{1}{8}x^2$; (c) $y = -5x^2$; (d) $y = -\frac{1}{12}x^2$.

22. A parabola has vertical axis and vertex at the origin. Write its equation if its focus is (a) $(0, 3)$; (b) $(0, 16)$; (c) $(0, -1)$; (d) $\left(0, -\frac{1}{10}\right)$.

23. Find the vertex and focus of each of the following parabolas, and state whether it opens up or down: (a) $y = x^2 - 4x + 1$; (b) $y = 2x^2 - 12x - 7$; (c) $y = -x^2 - 4x + 5$; (d) $y = 4 - 2x - \frac{1}{2}x^2$.

## 2.6 MORE ABOUT FUNCTIONS AND THEIR GRAPHS

In this section we sketch the graphs of a few of the simpler functions and comment briefly on the main features of these graphs.

**Linear functions.** These are functions of the form

$$y = f(x) = ax + b.$$

We recognize this as the equation of a straight line with slope $a$ and $y$-intercept $b$ (Fig. 17, left). If $a = 0$, as on the right, we speak of a *constant function*.

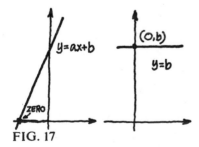

FIG. 17

For any function $y = f(x)$, a value of $x$ for which $f(x) = 0$ is called a *zero* of the function; the zeros of this function are thus the $x$-intercepts of its graph, or equivalently, the roots of the equation $f(x) = 0$. From the geometric point of view, the problem of solving a linear equation as discussed in Section 1.8 is simply the problem of finding the zero of a nonconstant linear function, as indicated in the figure.

**Quadratic functions.** These are functions of the form

$$y = f(x) = ax^2 + bx + c \qquad (a \neq 0).$$

If $a > 0$, we know that the graph of this function is a parabola that opens up. If the graph is low enough, then the function has two distinct zeros $r_1$ and $r_2$, as shown in Fig. 18. These are roots of the quadratic equation

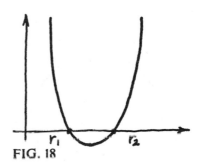

FIG. 18

$$ax^2 + bx + c = 0 \quad \text{or} \quad x^2 + \frac{b}{a}x + \frac{c}{a} = 0,$$

and can always be found by the quadratic formula. In terms of its roots, the second of these equations can be written in factored form as

$$(x - r_1)(x - r_2) = 0,$$

which multiplied out becomes

$$x^2 - (r_1 + r_2)x + r_1 r_2 = 0.$$

Accordingly, the sum and product of the roots are given by

$$r_1 + r_2 = -\frac{b}{a} \quad \text{and} \quad r_1 r_2 = \frac{c}{a}.$$

The cases of two equal real roots and of two distinct imaginary roots, as mentioned in Section 1.8, are shown in Fig. 19. In terms of the quadratic formula

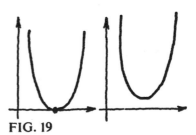

FIG. 19

$$x = \frac{-b \pm \sqrt{b^2 - 4ac}}{2a},$$

these cases correspond to the conditions $b^2 - 4ac = 0$ and $b^2 - 4ac < 0$.

Polynomial functions of higher degree can be difficult to sketch. However, in Fig. 20 we give the graph of the power function

$$y = f(x) = x^n$$

when $n$ is even and $\geq 2$ (on the left) and when $n$ is odd and $\geq 3$ (on the right). For larger values of

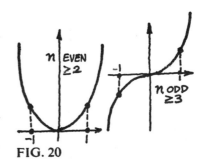

FIG. 20

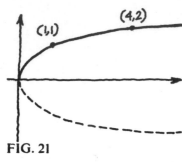

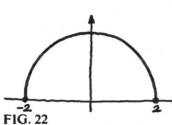

FIG. 21

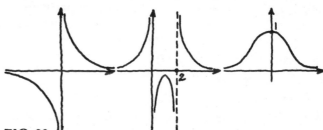

FIG. 22

$n$ these curves are flatter near $x = 0$ and steeper for $|x| > 1$.

**Radical functions.** The graph of the square root function

$$y = \sqrt{x}$$

is sketched in Fig. 21. This curve is the top half of the graph of the equation $y^2 = x$, which is a parabola opening to the right.

The graph of

$$y = \sqrt{4 - x^2}$$

is shown in Fig. 22; it is the top half of the circle $x^2 + y^2 = 4$.

**Rational functions.** Just as rational numbers are essentially quotients of integers, *rational functions* are quotients of polynomials. In Fig. 23 we sketch the graphs of the following three rational functions:

$$y = \frac{1}{x}, \qquad y = \frac{1}{x^2 - 2x} = \frac{1}{x(x-2)}, \qquad y = \frac{1}{x^2 + 1}.$$

FIG. 23

If the student examines these graphs carefully, he will see how important it is to pay special attention to the zeros of the denominator of a rational function.

One more word on functions and their graphs. Most elementary textbooks on algebra recommend sketching the graph of a function by computing a table of values and plotting many points. Plotting a few points can be useful, and is handy to have as a method of last resort when all else fails. However, a better approach is to examine the function intelligently and perceive its most characteristic features, and then to use these perceptions in sketching the graph.

## EXERCISES

24. Without actually finding the roots, determine for each equation whether its roots are

real and distinct, real and equal, or imaginary: (a) $5x^2 + 3x + 4 = 0$; (b) $7x^2 - 2x - 15 = 0$; (c) $3x^2 - 5x + 1 = 0$; (d) $4x^2 + 20x + 25 = 0$.

25. Without solving, find the sum and product of the roots: (a) $4x^2 - 7x - 13 = 0$; (b) $3x^2 + 10x + 17 = 0$; (c) $2x^2 + x - 2 = 0$.

26. Construct a quadratic equation having the given numbers as roots: (a) $-3$, $8$; (b) $2 + \sqrt{5}$, $2 - \sqrt{5}$; (c) $-\frac{3}{5}$, $\frac{2}{3}$.

27. Sketch the graph of $y = f(x)$ if $f(x)$ equals (a) $x^2 - 4x + 4$; (b) $x^2 - x - 2$; (c) $\frac{1}{x^2}$; (d) $\frac{x}{x^2 + 1}$.

# 3. SPECIAL TOPICS

## 3.1 LOGARITHMS

Logarithms are exponents. Thus, in the equation $100 = 10^2$, the exponent 2 is the logarithm of 100 to the base 10.

As a means of clarifying the concept of logarithms, we begin by pointing out that the equations

$$x = 2y \quad \text{and} \quad y = \frac{1}{2}x$$

express the same relation between $x$ and $y$, and in fact are really only one equation, written first in a form solved for $x$ and second in a form solved for $y$. In the same sense, if $a$ is any constant $> 1$, then the equations

$$x = a^y \quad \text{and} \quad y = \log_a x$$

are fully equivalent, except that the second is solved for $y$ and the symbol "$\log_a$" is invented to express this idea. As examples, we can write

| Statements about exponents | Equivalent statements about logarithms |
|---|---|
| $1000 = 10^3$ | $3 = \log_{10} 1000$ |
| $16 = 2^4$ | $4 = \log_2 16$ |
| $2 = 8^{1/3}$ | $\frac{1}{3} = \log_8 2$ |
| $\frac{1}{9} = 3^{-2}$ | $-2 = \log_3 \frac{1}{9}$ |
| $1 = a^0$ | $0 = \log_a 1$ |

The rules governing the behavior of logarithms are direct translations of corresponding rules of exponents.

| *Rules of logarithms* | *Proofs* |
|---|---|
| (1) $\log_a x_1 x_2 = \log_a x_1 + \log_a x_2$ | $x_1 x_2 = a^{y_1} \cdot a^{y_2}$ <br> $\qquad = a^{y_1 + y_2}$ |
| (2) $\log_a \dfrac{x_1}{x_2} = \log_a x_1 - \log_a x_2$ | $\dfrac{x_1}{x_2} = \dfrac{a^{y_1}}{a^{y_2}} = a^{y_1 - y_2}$ |
| (3) $\log_a x^n = n \log_a x$ | $x^n = (a^y)^n = a^{ny}$ |

In a bit more detail, the argument for rule (1) is as follows. If

$x_1 = a^{y_1}$ and $x_2 = a^{y_2}$, or equivalently,

$y_1 = \log_a x_1$ and $y_2 = \log_a x_2$,

then by a familiar rule of exponents,

$x_1 x_2 = a^{y_1} \cdot a^{y_2} = a^{y_1 + y_2}$.

But the exponent on the right, which is $\log_a x_1 x_2$, is also $\log_a x_1 + \log_a x_2$, and this is (1). The other two properties are proved similarly.

The student may find it illuminating to consider the following two useful facts and their verbal equivalents:

$a^{\log_a x} = x$:     when $x$ is expressed as some power of $a$, the exponent *is* $\log_a x$;

$\log_a a^x = x$:     the left side *is* the exponent when $a^x$ is expressed as some power of $a$.

We also point out that the equation

$x = b^y = (a^{\log_a b})^y = a^{y \log_a b}$

translates into

$\log_a x = (\log_a b)(\log_b x)$.

This is a convenient fact to have available for shifting logarithms from one base to another.

Logarithms to the base 10 were once widely used for numerical calculations, but this use has almost disappeared in these days of computers and hand calculators. Nevertheless, logarithm functions remain indispensable in higher mathematics and its applications to science.

# EXERCISES

1. Express each of the following in terms of logarithms: (a) $5^2 = 25$; (b) $2^5 = 32$; (c) $5^{-2} = \frac{1}{25}$; (d) $2^6 = 64$; (e) $3^3 = 27$; (f) $81^{0.5} = 9$; (g) $7^0 = 1$; (h) $10^{-1} = \frac{1}{10}$; (i) $32^{4/5} = 16$; (j) $16^{0.75} = 8$.

2. Express each of the following in terms of exponents: (a) $\log_{10} 10 = 1$; (b) $\log_4 8 = \frac{3}{2}$; (c) $\log_2 8 = 3$; (d) $\log_5 \frac{1}{125} = -3$; (e) $\log_9 81 = 2$; (f) $\log_{10} 1 = 0$; (g) $\log_{10} .01 = -2$; (h) $\log_7 343 = 3$; (i) $\log_5 125 = 3$; (j) $\log_{10} 0.1 = -1$.

3. Evaluate: (a) $\log_2 2$; (b) $\log_{10} 10{,}000$; (c) $\log_2 16$; (d) $\log_{25} 125$; (e) $\log_{10} .001$; (f) $\log_8 4$; (g) $\log_2 1$; (h) $\log_8 \dfrac{\sqrt{2} \cdot \sqrt[3]{256}}{\sqrt[6]{32}}$; (i) $\log_9 \dfrac{81(9)^{4/3}}{\sqrt[3]{9^7} \cdot \sqrt{9^0}}$.

4. Solve for $x$: (a) $\log_9 x = 3.5$; (b) $\log_{27} x = \frac{5}{3}$; (c) $\log_2 x = 8$; (d) $\log_{32} x = .8$.

5. Find the base $a$: (a) $\log_a 9 = .4$; (b) $\log_a 27 = -\frac{3}{4}$; (c) $\log_a 49 = 2$; (d) $\log_a 6 = \frac{1}{2}$.

# 3.2 DIVISION OF POLYNOMIALS

In arithmetic, division is often applied to reduce an "improper" fraction to the sum of an integer and a "proper" fraction, as in

$$\frac{14}{3} = 4 + \frac{2}{3}. \tag{1}$$

In algebra, division of polynomials is applied in a similar way to a rational function (quotient of polynomials) if the degree of the numerator is greater than or equal to the degree of the denominator. For example,

$$\frac{2x^3 + 3x^2 + 7x - 1}{x^2 + 1} = 2x + 3 + \frac{5x - 4}{x^2 + 1}. \tag{2}$$

In effect, the "improper" (top-heavy) rational function on the left is reduced to a polynomial plus a "proper" (bottom-heavy) rational function with the same denominator as the given rational function.

The structural pattern of (2) is this: if we are given a rational function $\frac{f(x)}{g(x)}$ in which $\deg f(x) \geq \deg g(x)$ ["deg" means degree], then dividing $f(x)$ by $g(x)$ means finding polynomials $Q(x)$ and $R(x)$ such that

$$\frac{f(x)}{g(x)} = Q(x) + \frac{R(x)}{g(x)}, \tag{3}$$

where $\deg R(x) < \deg g(x)$. It is customary to call $f(x)$ the *dividend*, $g(x)$ the *divisor*, $Q(x)$ the *quotient*, and $R(x)$ the *remainder*.

To understand how such a division is carried out, we consider how it is done in the case of (1), which is equivalent to

$$14 = 4 \cdot 3 + 2 \quad \text{or} \quad 14 - 4 \cdot 3 = 2.$$

The last equation displays the procedure: we subtract from 14 (the dividend) a multiple of 3 (the divisor) that leaves a remainder of 2, which is as small as possible without being negative. In much the same way, (2) is equivalent to

$$(2x^3 + 3x^2 + 7x - 1) . - (2x + 3)(x^2 + 1) = 5x - 4.$$

This equation gives a clue to the process of finding (2), which is basically the process of finding the quotient $2x + 3$. We see that the first term $2x^3$ of the dividend is the product of $2x$, the first term of the quotient, and $x^2$, the first term of the divisor. Hence the first term of the quotient can be found by dividing the first term of the dividend by the first term of the divisor, where in each polynomial the terms are written in order of decreasing exponents. The next step is to multiply the divisor by the first term of the quotient which we have just found, and then to subtract the result from the dividend, which reduces the degree of the remaining polynomial. This process is then repeated over and over until the residual polynomial has degree less than that of the divisor, at which point the procedure stops. It is convenient to arrange this work as follows:

$$\frac{2x + 3 \qquad \text{(quotient)}}{\text{(divisor)} \quad x^2 + 1 \overline{)\, 2x^3 + 3x^2 + 7x - 1 \quad \text{(dividend)}}}$$

$$\begin{array}{r} \underline{2x^3 \qquad + 2x} \\ 3x^2 + 5x - 1 \\ \underline{3x^2 \qquad + 3} \\ 5x - 4 \quad \text{(remainder)} \end{array}$$

The method described here is called *long division*. The division indicated in (3) can be written in the form

$$f(x) = Q(x)g(x) + R(x) \quad \text{or}$$
$$f(x) - Q(x)g(x) = R(x),$$

and in any specific case can be carried out in just the same way.

**Example.** To divide $2x^3 + x^2 - 13x + 10$ by $x - 2$, we write

$$\begin{array}{r} 2x^2 + 5x \ - 3 \\ x - 2 \overline{)\, 2x^3 + \ x^2 - 13x + 10} \\ \underline{2x^3 - 4x^2} \\ 5x^2 - 13x + 10 \\ \underline{5x^2 - 10x} \\ -3x + 10 \\ \underline{-3x + \ 6} \\ 4 \end{array}$$

This result can be written as

$$\frac{2x^3 + x^2 - 13x + 10}{x - 2} = 2x^2 + 5x - 3 + \frac{4}{x - 2},$$

or as

$$2x^3 + x^2 - 13x + 10 = (2x^2 + 5x - 3)(x - 2) + 4.$$

If a polynomial $f(x)$ is divided by a linear polynomial $x - r$, then the quotient is a polynomial $Q(x)$ and the remainder is a constant $R$,

$$f(x) = (x - r)Q(x) + R.$$

The value of this constant is easily discovered by putting $x = r$, which yields $f(r) = 0 + R$ or $R = f(r)$. This says that the remainder when $f(x)$ is divided by $x - r$ is the value of $f(x)$ at $x = r$. Summarizing, we have the

**Remainder Theorem.** If $f(x)$ is a polynomial and $r$ is any real number, then the remainder when $f(x)$ is divided by $x - r$ is $f(r)$; that is,

$$f(x) = (x - r)Q(x) + f(r)$$

where $Q(x)$ is a polynomial.

**67**

In particular, we see from this that $f(r) = 0$ if and only if $f(x) = (x - r)Q(x)$, that is, if and only if $x - r$ is a factor of $f(x)$. This conclusion is the important

**Factor Theorem.** If $f(x)$ is a polynomial, then a real number $r$ is a root of the equation $f(x) = 0$ if and only if $x - r$ is a factor of $f(x)$.

## EXERCISES

6. In each of the following, use long division to find the quotient $Q(x)$ and remainder $R(x)$:

   (a) $\dfrac{x^2 - 5x + 6}{x - 3}$;

   (b) $\dfrac{x^2 - 4x - 21}{x - 7}$;

   (c) $\dfrac{x^3 - 8x^2 + x + 42}{x - 7}$;

   (d) $\dfrac{3x^4 - 2x^3 + 10x^2 - 7x + 10}{x^2 + 3}$;

   (e) $\dfrac{x^5 - 4x^4 + 3x^3 + 7x^2 - 10x - 5}{x^2 - 2x - 1}$.

7. Use the factor theorem to show that (a) $x^{91} + 3x^{73} - 2x^{37} - 2$ has $x - 1$ as a factor; (b) $x^{91} + x^{74} + x^{132} + x^{51}$ has $x + 1$ as a factor; (c) $x^5 - 3x^4 + 6x^3 - 5x^2 - 12$ has $x - 2$ as a factor.

8. If $n$ is any positive integer, use the factor theorem to show that $x^n - 1$ has $x - 1$ as a factor, and then use long division to compute the other factor.

9. Factor completely each of the following polynomials $f(x)$ by noticing that the given number is a root of the equation $f(x) = 0$:
   (a) $x^3 - 6x^2 + 11x - 6$, $x = 1$; (b) $x^3 - x^2 + x - 1$, $x = 1$; (c) $x^3 + 5x^2 - 12x - 36$, $x = -2$; (d) $x^3 - x^2 - 5x - 3$, $x = 3$.

## 3.3 DETERMINANTS AND SYSTEMS OF LINEAR EQUATIONS

The symbol

$$\begin{vmatrix} a_1 & b_1 \\ a_2 & b_2 \end{vmatrix} \tag{1}$$

is called a *second order determinant*. It is used

to denote the number $a_1b_2 - b_1a_2$; that is, by definition we have

$$\begin{vmatrix} a_1 & b_1 \\ a_2 & b_2 \end{vmatrix} = a_1b_2 - b_1a_2. \tag{2}$$

The number $a_1b_2 - b_1a_2$ is called the *value* or *expansion* of the determinant (1), and the four numbers $a_1$, $b_1$, $a_2$, $b_2$ are called its *elements*.

**Example.**

$$\begin{vmatrix} 2 & -5 \\ -3 & 6 \end{vmatrix} = 2 \cdot 6 - (-5) \cdot (-3)$$

$$= 12 - 15 = -3.$$

We point out that the value of our second order determinant is most easily remembered as the product of the two elements in the diagonal from upper left to lower right, minus the product of the two elements in the diagonal from upper right to lower left. This rule is suggested in the following diagram:

We now use this determinant notation to derive a convenient set of formulas for the solution of a system of two linear equations in two unknowns. Consider the system

$$a_1x + b_1y = c_1,$$
$$a_2x + b_2y = c_2. \tag{3}$$

To solve this system, we first eliminate $y$ by multiplying the equations by $b_2$ and $b_1$, respectively, and subtracting; we next eliminate $x$ by multiplying the equations by $a_2$ and $a_1$, respectively, and subtracting; and we write down the result of these procedures:

$$(a_1b_2 - b_1a_2)x = c_1b_2 - b_1c_2,$$
$$(b_1a_2 - a_1b_2)y = c_1a_2 - a_1c_2.$$

If the number $a_1b_2 - b_1a_2 \neq 0$, we can solve these two equations for $x$ and $y$:

$$x = \frac{c_1b_2 - b_1c_2}{a_1b_2 - b_1a_2}, \qquad y = \frac{a_1c_2 - c_1a_2}{a_1b_2 - b_1a_2}. \tag{4}$$

It will be noticed that the denominators in (4) are simply the value of the determinant (1), and that the numerators for $x$ and $y$ are the values of

$$\begin{vmatrix} c_1 & b_1 \\ c_2 & b_2 \end{vmatrix} \quad \text{and} \quad \begin{vmatrix} a_1 & c_1 \\ a_2 & c_2 \end{vmatrix}.$$

We can therefore write equations (4) in the form

$$x = \frac{\begin{vmatrix} c_1 & b_1 \\ c_2 & b_2 \end{vmatrix}}{\begin{vmatrix} a_1 & b_1 \\ a_2 & b_2 \end{vmatrix}}, \qquad y = \frac{\begin{vmatrix} a_1 & c_1 \\ a_2 & c_2 \end{vmatrix}}{\begin{vmatrix} a_1 & b_1 \\ a_2 & b_2 \end{vmatrix}}. \tag{5}$$

In each case the denominator is the determinant of the array of coefficients on the left of (3). In the formula for $x$, the numerator is the determinant in the denominator with the column of $x$-coefficients replaced by the column of constants on the right of (3); and in the formula for $y$, the numerator has the column of $y$-coefficients replaced by the column of constants.

**Example.** To solve the system

$$\begin{aligned} 4y &= x - 2, \\ 3x &= 4 - y, \end{aligned} \tag{6}$$

we first rearrange the equations into the standard form (3):

$$\begin{aligned} x - 4y &= 2, \\ 3x + y &= 4. \end{aligned}$$

By (5), the solution of the system is

$$x = \frac{\begin{vmatrix} 2 & -4 \\ 4 & 1 \end{vmatrix}}{\begin{vmatrix} 1 & -4 \\ 3 & 1 \end{vmatrix}} = \frac{2 \cdot 1 - (-4) \cdot 4}{1 \cdot 1 - (-4) \cdot 3} = \frac{2 + 16}{1 + 12} = \frac{18}{13},$$

$$y = \frac{\begin{vmatrix} 1 & 2 \\ 3 & 4 \end{vmatrix}}{\begin{vmatrix} 1 & -4 \\ 3 & 1 \end{vmatrix}} = \frac{1 \cdot 4 - 2 \cdot 3}{13} = \frac{-2}{13} = -\frac{2}{13}.$$

In geometric language, this simultaneous solution of the system (6) gives the coordinates of the point in the plane at which the straight lines (6) intersect.

We next discuss third order determinants and their use in solving systems of three linear equations in three unknowns.

We define the value of a *third order determinant* by means of the elements in its first row and certain corresponding second order determinants, as follows:

$$\begin{vmatrix} a_1 & b_1 & c_1 \\ a_2 & b_2 & c_2 \\ a_3 & b_3 & c_3 \end{vmatrix} = a_1 \begin{vmatrix} b_2 & c_2 \\ b_3 & c_3 \end{vmatrix} - b_1 \begin{vmatrix} a_2 & c_2 \\ a_3 & c_3 \end{vmatrix} + c_1 \begin{vmatrix} a_2 & b_2 \\ a_3 & b_3 \end{vmatrix}.$$

(7)

This definition is bound to seem mysterious, and the mystery cannot be dispelled without an in-depth study of determinants. Nevertheless, we will explain as best we can.*

The second order determinants in (7) are called the *minors* of the elements $a_1$, $b_1$, $c_1$. In general, the minor of any element of the determinant on the left side of (7) is the second order determinant that remains after we delete the row and column containing the element in question. Thus, the minors of $a_1$ and $b_1$ are

$$\begin{vmatrix} a_1 & b_1 & c_1 \\ a_2 & b_2 & c_2 \\ a_3 & b_3 & c_3 \end{vmatrix} = \begin{vmatrix} b_2 & c_2 \\ b_3 & c_3 \end{vmatrix},$$

$$\begin{vmatrix} a_1 & b_1 & c_1 \\ a_2 & b_2 & c_2 \\ a_3 & b_3 & c_3 \end{vmatrix} = \begin{vmatrix} a_2 & c_2 \\ a_3 & c_3 \end{vmatrix}.$$

It is a remarkable fact—which we cannot explain briefly—that the value of a third order determinant as defined by (7) can also be found by multiplying the elements of *any* row (or column) by their respective minors, with signs attached to the terms according to this pattern:

$$\begin{vmatrix} + & - & + \\ - & + & - \\ + & - & + \end{vmatrix}.$$

From this point of view, the definition (7) can now be regarded as one among many possible expansions, by means of the first row. The expansion by the second column is

---

*The role of determinants in elementary algebra is a good illustration of Marshall's well-known Generalized Iceberg Theorem: seven-eights of everything can't be seen.

$$\begin{vmatrix} a_1 & b_1 & c_1 \\ a_2 & b_2 & c_2 \\ a_3 & b_3 & c_3 \end{vmatrix} = -b_1 \begin{vmatrix} a_2 & c_2 \\ a_3 & c_3 \end{vmatrix} + b_2 \begin{vmatrix} a_1 & c_1 \\ a_3 & c_3 \end{vmatrix}$$

$$- b_3 \begin{vmatrix} a_1 & c_1 \\ a_2 & c_2 \end{vmatrix}.$$

The theory of determinants (which we naturally omit) guarantees that each method of expansion described above yields the same result.

**Example.** We calculate the value of the following determinant by using the first row, and then by using the second column:

$$\begin{vmatrix} 1 & -2 & 4 \\ 3 & -1 & 6 \\ 2 & 3 & 2 \end{vmatrix} = 1 \cdot \begin{vmatrix} -1 & 6 \\ 3 & 2 \end{vmatrix} - (-2) \cdot \begin{vmatrix} 3 & 6 \\ 2 & 2 \end{vmatrix}$$

$$+ 4 \cdot \begin{vmatrix} 3 & -1 \\ 2 & 3 \end{vmatrix} = (-2 - 18)$$

$$+ 2(6 - 12) + 4(9 + 2) = 12;$$

$$\begin{vmatrix} 1 & -2 & 4 \\ 3 & -1 & 6 \\ 2 & 3 & 2 \end{vmatrix} = -(-2) \cdot \begin{vmatrix} 3 & 6 \\ 2 & 2 \end{vmatrix} + (-1) \cdot \begin{vmatrix} 1 & 4 \\ 2 & 2 \end{vmatrix}$$

$$- 3 \cdot \begin{vmatrix} 1 & 4 \\ 3 & 6 \end{vmatrix} = 2(6 - 12)$$

$$- (2 - 8) - 3(6 - 12) = 12.$$

At this point we merely state the facts, omitting all computational details. Consider the following system of three linear equations in three unknowns:

$$a_1 x + b_1 y + c_1 z = d_1,$$
$$a_2 x + b_2 y + c_2 z = d_2, \tag{8}$$
$$a_3 x + b_3 y + c_3 z = d_3.$$

If this system is solved by standard algebraic methods of elimination, its solution will be found to be expressible by means of determinants in the form

$$x = \frac{\begin{vmatrix} d_1 & b_1 & c_1 \\ d_2 & b_2 & c_2 \\ d_3 & b_3 & c_3 \\ a_1 & b_1 & c_1 \\ a_2 & b_2 & c_2 \\ a_3 & b_3 & c_3 \end{vmatrix}}{\begin{vmatrix} a_1 & b_1 & c_1 \\ a_2 & b_2 & c_2 \\ a_3 & b_3 & c_3 \end{vmatrix}}, \qquad y = \frac{\begin{vmatrix} a_1 & d_1 & c_1 \\ a_2 & d_2 & c_2 \\ a_3 & d_3 & c_3 \end{vmatrix}}{\begin{vmatrix} a_1 & b_1 & c_1 \\ a_2 & b_2 & c_2 \\ a_3 & b_3 & c_3 \end{vmatrix}},$$

$$z = \frac{\begin{vmatrix} a_1 & b_1 & d_1 \\ a_2 & b_2 & d_2 \\ a_3 & b_3 & d_3 \end{vmatrix}}{\begin{vmatrix} a_1 & b_1 & c_1 \\ a_2 & b_2 & c_2 \\ a_3 & b_3 & c_3 \end{vmatrix}}, \tag{9}$$

provided the determinant in the denominators is nonzero. As before, we point out that the numerator of the expression for $x$ differs from the denominator only in that each $a$ is replaced by the corresponding $d$; the numerator for $y$, in that each $b$ is replaced by the corresponding $d$; and the numerator for $z$, in that each $c$ is replaced by the corresponding $d$.

The formulas (5) and (9) giving solutions of the systems (3) and (8) are usually called *Cramer's rule*. In the advanced theory of determinants it is shown that Cramer's rule works in exactly the same way for a system of $n$ linear equations in $n$ unknowns, where $n$ is any positive integer.

## EXERCISES

10.  Evaluate: (a) $\begin{vmatrix} 1 & 2 \\ 3 & 4 \end{vmatrix}$; (b) $\begin{vmatrix} -2 & 7 \\ -5 & -6 \end{vmatrix}$.

11.  Evaluate:

(a) $\begin{vmatrix} 2 & 0 & -1 \\ 3 & 2 & 6 \\ -4 & 5 & 0 \end{vmatrix}$;

(b) $\begin{vmatrix} -2 & 4 & 6 \\ 3 & 0 & 1 \\ 1 & 1 & -7 \end{vmatrix}$.

12. Solve by determinants:
  (a) $6x + 7y = 18$,
  $9x - 2y = -48$;
  (b) $2x + 3y + z = 4$,
  $x + 5y - 2z = -1$,
  $3x - 4y + 4z = -1$.

# 3.4 GEOMETRIC PROGRESSIONS AND SERIES

A *geometric progression* is a sequence of numbers, with a first, a second, a third, and so on, in which each number after the first is obtained from the one that precedes it by multiplying by a fixed number called the *ratio:*

$$a, ar, ar^2, ar^3, \ldots, ar^n, \ldots, \quad r = \text{the ratio.} \quad (1)$$

**Examples.** The following are geometric progressions:

| | | | | | |
|---|---|---|---|---|---|
| 2, | 4, | 8, | 16, | $\ldots$, | $r = 2$; |
| 2, | 6, | 18, | 54, | $\ldots$, | $r = 3$; |
| 1, | $\frac{1}{2}$, | $\frac{1}{4}$, | $\frac{1}{8}$, | $\ldots$, | $r = \frac{1}{2}$; |
| 2, | $-\frac{2}{3}$, | $\frac{2}{9}$, | $-\frac{2}{27}$, | $\ldots$, | $r = -\frac{1}{3}$. |

The basic fact about the progression (1) is formula (3) below, which gives the sum of its first $n + 1$ terms: if $S$ denotes the sum

$$S = a + ar + ar^2 + \cdots + ar^n, \quad (2)$$

then

$$S = \frac{a(1 - r^{n+1})}{1 - r}. \quad (3)$$

This formula clearly has no meaning if $r = 1$, so we assume that $r \neq 1$. To establish (3), multiply (2) through by $r$, obtaining

$$Sr = ar + ar^2 + ar^3 + \cdots + ar^{n+1}, \quad (4)$$

and then subtract (4) from (2) by exploiting all possible cancellations. We write (2) and (4) together to emphasize these cancellations,

$$S = a + ar + ar^2 + \cdots + ar^n,$$
$$Sr = ar + ar^2 + ar^3 + \cdots + ar^{n+1}.$$

It is now easy to see that

$$S - Sr = a - ar^{n+1} \quad \text{or}$$
$$S(1 - r) = a(1 - r^{n+1}),$$

which is equivalent to (3).

If the sum in (2) is extended to an infinite number of terms, this is indicated in symbols by using dots in the following way:

$$a + ar + ar^2 + \cdots + ar^n + \cdots. \tag{5}$$

This sum is called a *geometric series*. Its value — if it has one — is determined by examining the behavior of the *partial sum*

$$S = a + ar + ar^2 + \cdots + ar^n$$

as the positive integer $n$ increases. By formula (3), this partial sum can be written in the form

$$S = \frac{a(1 - r^{n+1})}{1 - r} = \frac{a}{1 - r} - \frac{a}{1 - r}r^{n+1}. \tag{6}$$

If the ratio $r$ is any number numerically less than 1, that is, if $|r| < 1$, then the second term on the right of (6) clearly approaches zero as $n$ increases. (If a number numerically less than 1 is raised to higher and higher powers, it gets smaller and smaller.) This can be expressed by writing

$$\frac{a}{1 - r}r^{n+1} \to 0 \quad \text{as } n \to \infty,$$

where the arrow means "approaches." For these values of $r$ we therefore have

$$S = a + ar + ar^2 + \cdots + ar^n \to \frac{a}{1 - r} \quad \text{as } n \to \infty.$$

This is what is meant by the statement that the formula

$$a + ar + ar^2 + \cdots + ar^n + \cdots = \frac{a}{1 - r} \tag{7}$$

is valid for $|r| < 1$.

**Example.** Formula (7) can be used to show that any infinite repeating decimal represents a rational number. For instance, we have the familiar fact that

$$.333\ldots = \frac{1}{3}.$$

To understand why this is true, we use the meaning of the decimal — and apply formula (7) at the right moment — to write

$$.333\ldots = \frac{3}{10} + \frac{3}{10^2} + \frac{3}{10^3} + \cdots$$

$$= \frac{3}{10} + \left(\frac{3}{10}\right)\left(\frac{1}{10}\right) + \left(\frac{3}{10}\right)\left(\frac{1}{10}\right)^2 + \cdots$$

$$\frac{\frac{3}{10}}{1 - \frac{1}{10}} = \frac{3}{10} \cdot \frac{10}{9} = \frac{1}{3}.$$

# EXERCISES

13. If 5 bacteria enter a human body at the same time, and if each divides in two every 12 hours and none die, how many will there be a week later? Hint: after 1 day there will be $5 \cdot 2 \cdot 2 = 5 \cdot 4$.

14. A golf ball is dropped from a height of 81 inches. If it always rebounds $\frac{2}{3}$ of the distance it falls, use formula (3) to find the total distance it has traveled if it is caught at the top of the fourth bounce.

15. Find each of the following sums:

    (a) $1 + \frac{1}{2} + \frac{1}{4} + \cdots$ ;

    (b) $4 - 2 + 1 - \cdots$ ;

    (c) $9 + 6 + 4 + \cdots$ ;

    (d) $6 - 2 + \frac{2}{3} - \cdots$ ;

    (e) $3 + \sqrt{3} + 1 + \cdots$ ;

    (f) $\sqrt{12} + \sqrt{6} + \sqrt{3} + \cdots$ .

16. Express as a fraction:

    (a) $.777\ldots$ ;

    (b) $.343434\ldots$ ;

    (c) $3.72444\ldots$ .

17. What is the total distance traveled by the golf ball in Exercise 14 before it comes to rest?

18. In an infinite nested sequence of equilateral triangles, the vertices of each triangle after the first are the midpoints of the sides of the preceding triangle. Find the sum of the perimeters of all the triangles if the perimeter of the first triangle is 12 inches.

## 3.5 ARITHMETIC PROGRESSIONS

An *arithmetic progression* is a sequence of numbers in which each term after the first is obtained from the one that precedes it by adding a fixed number called the *common difference:*

$$a, \quad a+d, \quad a+2d, \quad \ldots, \quad a+(n-1)d. \quad (1)$$

The simplest progression of this type consists of the first $n$ positive integers,

$$1, \quad 2, \quad 3, \quad \ldots, \quad n.$$

This is also the most important arithmetic progression, for if $S$ denotes its sum, that is, if

$$S = 1 + 2 + 3 + \cdots + n,$$

then a formula for $S$ is necessary in many applications. We can easily find this formula by writing the sum twice, once as given and the second time in reverse order:

$$S = 1 + \quad 2 \quad + \quad 3 \quad + \cdots + n,$$
$$S = n + (n-1) + (n-2) + \cdots + 1.$$

By adding, we get $2S$ on the left; and since each column on the right adds up to $n+1$ and there are $n$ columns, we have

$$2S = n(n+1)$$

or

$$S = \frac{n(n+1)}{2},$$

and this is the formula mentioned above.

## EXERCISES

19. If the first term and the $n$th term of the general arithmetic progression (1) are denoted by $a_1$ and $a_n$, then by working in from the ends its sum can be written in the form $S = a_1 + (a_1 + d) + \cdots + (a_n - d) + a_n$. Use the reversing device employed above to show that $S = n\left(\frac{a_1 + a_n}{2}\right)$. Notice that the quantity in parentheses is the average of the first and last terms.

20. Use the result of the preceding exercise to find a formula for the sum of the first $n$ odd numbers. Hint: how can the $n$th odd number be expressed in terms of $n$?

## 3.6 PERMUTATIONS AND COMBINATIONS

Our subject here is certain techniques of counting that are useful in a variety of situations, in particular, in connection with the binomial theorem discussed in the next section.

We begin by describing a special notation. If $n$ is a positive integer, the product of all the positive integers up to $n$ is $1 \cdot 2 \cdot 3 \cdots n$. This product is customarily denoted by $n!$ and called "$n$ factorial":

$$n! = 1 \cdot 2 \cdot 3 \cdots n.$$

Thus, $1! = 1$, $2! = 1 \cdot 2 = 2$, $3! = 1 \cdot 2 \cdot 3 = 6$, $4! = 1 \cdot 2 \cdot 3 \cdot 4 = 24$, and so on. For technical reasons, we define $0!$ to be 1. The factorials increase very rapidly, as we see by doing a little arithmetic:

$$5! = 120, \qquad 6! = 720, \qquad 7! = 5040,$$
$$8! = 40,320, \qquad 9! = 362,880, \qquad 10! = 3,628,800.$$

The reasoning on which our work with permutations is based can be illustrated by a simple example. Consider a journey from a city $A$ through a city $B$ to a city $C$. Suppose it is possible to go from $A$ to $B$ by 4 different routes and from $B$ to $C$ by 3 different routes. Then the total number of different routes from $A$ through $B$ to $C$ is $4 \cdot 3 = 12$; for we can go from $A$ to $B$ in any one of the 4 ways, and for each of these ways there are 3 ways of going on from $B$ to $C$.

The basic principle here is this: if two successive independent decisions are to be made, and if there are $c_1$ choices for the first and $c_2$ choices for the second, then the total number of ways of making these two decisions is the product $c_1 c_2$. It is clear that the same principle is valid for any number of successive independent decisions.

The following is our main application of this idea. Given $n$ distinct objects, in how many ways can we arrange them in order, that is, with a first, a second, a third, and so on? The answer is easy. There are $n$ choices for the first object. After the first object is chosen, there are $n - 1$ choices for the second, then $n - 2$ choices for the third, and so on. By the basic principle stated above, the total number of orderings is therefore

$$n(n-1)(n-2) \cdots 2 \cdot 1 = n!.$$

Each ordering of a set of objects is called a *permutation* of those objects. We have reached the following conclusion:

the number of permutations of $n$ objects is $n!$.

**Examples.** (a) There are $6! = 720$ ways of arranging 6 books on a shelf. (b) Since there are 9 players on a baseball team, there are $9! = 362,880$ possible batting orders for these players.

We next consider a slight generalization. Suppose again that we have $n$ distinct objects. This time we ask how many ways $k$ of them can be chosen in order. Each such ordering is called a *permutation of* n *objects taken* k *at a time*, and the total number of these permutations is denoted by $P(n, k)$. There are evidently $n$ choices for the first, $n-1$ choices for the second, $n-2$ choices for the third, and $n-(k-1) = n-k+1$ choices for the $k$th. The total number of these permutations is therefore

$$P(n, k) = n(n-1)(n-2) \cdots (n-k+1).$$

If we write this number in terms of factorials, by inserting additional factors and then cancelling them out again, we can state our conclusion as follows:

the number of permutations of $n$ objects taken $k$ at a time is $P(n, k) = n(n-1)(n-2) \cdots$
$(n-k+1) = \dfrac{n!}{(n-k)!}$.

**Examples.** (a) If we have 6 books and only 3 spaces available on a bookshelf, then the number of ways of filling these spaces with the books on hand (counting the order of the books as they stand on the shelf) is

$$P(6, 3) = \frac{6!}{(6-3)!} = \frac{6!}{3!} = 6 \cdot 5 \cdot 4 = 120.$$

(b) The number of ways (counting the order of the cards) in which a 5-card poker hand can be dealt from a deck of 52 cards is

$$P(52, 5) = \frac{52!}{(52-5)!} = \frac{52!}{47!} = 52 \cdot 51 \cdot 50 \cdot 49 \cdot 48$$
$$= 311,875,200.$$

Of course, the order of the cards in a poker hand is immaterial to the value of the hand, so the number of distinct poker hands is a considerably smaller number. We take account of this below, in our discussion of combinations.

A set of $k$ objects chosen from a given set of $n$ objects, without regard to the order in which they are arranged, is called a *combination of* n *objects taken* k *at a time*. The total number of such combinations is sometimes denoted by $C(n, k)$, but more frequently by $\binom{n}{k}$. For reasons explained in the next section, the numbers $\binom{n}{k}$ are called *binomial coefficients*.

Permutations and combinations are related in a simple way. Each permutation of $n$ objects taken $k$ at a time consists of a choice of $k$ objects (a combination) followed by an ordering of these $k$ objects. But there are $\binom{n}{k}$ ways to choose $k$ objects, and then $k!$ ways to arrange them in order, so

$$P(n, k) = \binom{n}{k} \cdot k! \quad \text{or} \quad \binom{n}{k} = \frac{P(n, k)}{k!}.$$

Our formula for $P(n, k)$ now yields the following conclusion:

the number of combinations of $n$ objects taken $k$ at a time is $\binom{n}{k} = \dfrac{n!}{k!(n-k)!}$.

**Examples.** (a) The number of committees of 4 people that can be chosen from a group of 7 people is

$$\binom{7}{4} = \frac{7!}{4!3!} \quad \frac{7 \cdot 6 \cdot 5}{2 \cdot 3} = 35.$$

(b) The number of different 5-card poker hands that can be dealt from a deck of 52 cards is

$$\binom{52}{5} = \frac{52!}{5!47!} = \frac{52 \cdot 51 \cdot 50 \cdot 49 \cdot 48}{2 \cdot 3 \cdot 4 \cdot 5} = 2,598,960.$$

## EXERCISES

21. Compute each of the following:

(a) $\dfrac{9!}{6!}$;

(b) $\dfrac{13!}{7!}$;

(c) $\dbinom{18}{3}$;

(d) $\dbinom{36}{4}$.

22. If 8 horses run in a race, how many different orders of finishing are there? How many possibilities are there for the first three places (win, place, and show)?

23. A club has 14 members. In how many ways can a president, a vice president, and a secretary be chosen?

24. In an examination a student has a choice of any 3 out of 9 questions. How many ways can he choose his questions?

25. In how many ways can a jury of 6 people be selected from a panel of 15 eligible citizens?

## 3.7 THE BINOMIAL THEOREM

The binomial theorem is a general formula for the expanded $n$-factor product

$$(a + b)^n = (a + b)(a + b) \cdots (a + b). \qquad (1)$$

If we compute a few cases by repeated laborious multiplication, we find that

$(a + b)^1 = a + b$,

$(a + b)^2 = a^2 + 2ab + b^2$,

$(a + b)^3 = a^3 + 3a^2b + 3ab^2 + b^3$,

$(a + b)^4 = a^4 + 4a^3b + 6a^2b^2 + 4ab^3 + b^4$,

$(a + b)^5 = a^5 + 5a^4b + 10a^3b^2 + 10a^2b^3$
$\qquad\qquad + 5ab^4 + b^5$.

It is clear that the corresponding expansion of (1) begins with $a^n$ and ends with $b^n$, and also that the intermediate terms involve steadily decreasing powers of $a$ and steadily increasing powers of $b$ so that the sum of the two exponents is exactly $n$ in each term. What is not so clear is the way the coefficients are calculated.

The general form of the expansion of (1) — the *binomial theorem* — is

$$(a + b)^n = a^n + na^{n-1}b + \frac{n(n-1)}{2}a^{n-2}b^2$$
$$+ \frac{n(n-1)(n-2)}{2 \cdot 3}a^{n-3}b^3 + \cdots$$
$$+ \frac{n(n-1)(n-2) \cdots (n-k+1)}{1 \cdot 2 \cdot 3 \cdots k}a^{n-k}b^k$$
$$+ \cdots + b^n. \tag{2}$$

To prove this, all that is necessary is to understand the reasons behind the form of these coefficients. This is not difficult if we look at (1) and observe that each term of the expansion can be thought of as the product of $n$ letters, one taken from each factor of the product

$$(a + b)(a + b) \cdots (a + b), \quad n \text{ factors.}$$

Thus, a product $a^{n-k}b^k$ is obtained by choosing $k$ $b$'s and the rest $a$'s. The number of ways this can be done is $\binom{n}{k}$, the number of combinations of $n$ objects taken $k$ at a time, so this is the number of times the term $a^{n-k}b^k$ occurs in the expansion. The coefficient of $a^{n-k}b^k$ on the right side of (2) must therefore be $\binom{n}{k}$, and since

$$\binom{n}{k} = \frac{n!}{k!(n-k)!}$$
$$= \frac{n(n-1)(n-2) \cdots (n-k+1)}{1 \cdot 2 \cdot 3 \cdots k},$$

the proof of (2) is complete.

**Example.** To use (2) in finding the expansion of $(a + b)^5$, we write

$$(a + b)^5 = a^5 + \binom{5}{1}a^4b + \binom{5}{2}a^3b^2 + \binom{5}{3}a^2b^3$$
$$+ \binom{5}{4}ab^4 + b^5$$
$$= a^5 + \frac{5!}{1!4!}a^4b + \frac{5!}{2!3!}a^3b^2 + \frac{5!}{3!2!}a^2b^3$$
$$+ \frac{5!}{4!1!}ab^4 + b^5$$
$$= a^5 + 5a^4b + \frac{5 \cdot 4}{2}a^3b^2 + \frac{5 \cdot 4 \cdot 3}{2 \cdot 3}a^2b^3$$
$$+ \frac{5 \cdot 4 \cdot 3 \cdot 2}{2 \cdot 3 \cdot 4}ab^4 + b^5$$
$$= a^5 + 5a^4b + 10a^3b^2 + 10a^2b^3 + 5ab^4$$
$$+ b^5.$$

## EXERCISE

26.  Write out the expansions of (a) $(a + b)^9$; (b) $(2a + b)^6$; (c) $(2a - 3b)^5$.

# 3.8 MATHEMATICAL INDUCTION

In Section 3.5 we established a formula for the sum of the first $n$ positive integers,

$$1 + 2 + 3 + \cdots + n = \frac{n(n + 1)}{2}.$$

It is also occasionally useful to know the corresponding formula for the sum of the first $n$ squares,

$$1^2 + 2^2 + 3^2 + \cdots + n^2 = \frac{n(n + 1)(2n + 1)}{6}. \quad (1)$$

How is such a formula discovered? There are several interesting ways to do this, but we do not discuss them here.

Formula (1) cannot be established by the argument used earlier, since the terms on the left do not form an arithmetic progression. Our present purpose is to give a brief description of a powerful method of proof that *will* work, and to illustrate this method by applying it to formula (1).

**The Principle of Mathematical Induction.** A statement $S(n)$ that depends on a positive integer $n$ is true for all $n$ if it satisfies the following two conditions:

(a) $S(n)$ is true for $n = 1$;

(b) if $S(n)$ is true for $n = k$, then it is also true for $n = k + 1$.

We will not discuss this principle in detail, or attempt to explain its intuitive content. However, we do offer the following equivalent version, which some students may find easier to understand and accept: any set of positive integers that contains 1, and also contains $k + 1$ whenever it contains $k$, necessarily contains all positive integers.

As our illustration, we take (1) to be the statement $S(n)$. Our first step is to verify that (1) is true for $n = 1$, which is easy:

$$1^2 = \frac{1 \cdot 2 \cdot 3}{6}.$$

Next, we assume that (1) is true for $n = k$,

$$1^2 + 2^2 + 3^2 + \cdots + k^2 = \frac{k(k + 1)(2k + 1)}{6}, \quad (2)$$

and then we try to prove (1) for $n = k + 1$:

$$1^2 + 2^2 + 3^2 + \cdots + k^2 + (k+1)^2$$
$$= \frac{(k+1)(k+2)(2k+3)}{6}. \tag{3}$$

But by using our hypothesis (2), the left side of (3) can be written as

$$1^2 + 2^2 + 3^2 + \cdots + k^2 + (k+1)^2$$
$$= \frac{k(k+1)(2k+1)}{6} + (k+1)^2$$
$$= (k+1)\left[\frac{k(2k+1)}{6} + k + 1\right]$$
$$= \frac{(k+1)(2k^2 + 7k + 6)}{6}$$
$$= \frac{(k+1)(k+2)(2k+3)}{6},$$

which is (3). We now appeal to the principle to conclude that (1) is true for all $n$.

## EXERCISE

27. Apply the principle of mathematical induction to prove that each of the following statements is true for all positive integers $n$:

    (a) $1 + 3 + 5 + \cdots + (2n - 1) = n^2$;

    (b) $1^2 + 3^2 + 5^2 + \cdots + (2n - 1)^2$
    $$= \frac{n(4n^2 - 1)}{3};$$

    (c) $1^3 + 2^3 + 3^3 + \cdots + n^3 = \left[\frac{n(n+1)}{2}\right]^2$;

    (d) $2^{2n} - 1$ is divisible by 3;

    (e) $3^{2n} - 1$ is divisible by 8.

## 3.9 THE CONE AND SPHERE REVISITED

In Sections 3.5 and 3.8 we proved the formulas

$$1 + 2 + 3 + \cdots + n = \frac{n(n+1)}{2} \tag{1}$$

and

$$1^2 + 2^2 + 3^2 + \cdots + n^2 = \frac{n(n+1)(2n+1)}{6} \tag{2}$$

for the sum of the first $n$ integers and the sum of the first $n$ squares. We remind the reader that the most complicated parts of Chapter 1 are those concerned with the formulas for the volume of a cone and a sphere, as shown in Fig. 24. Our purpose in this section is to demonstrate the value of formulas (1) and (2) by showing how they can be used to give new and totally different proofs of these important facts from geometry.

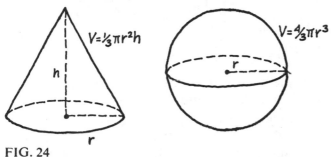

$V = \frac{1}{3}\pi r^2 h$

$V = \frac{4}{3}\pi r^3$

FIG. 24

First, the cone. We imagine that a large number of thin circular discs of equal thickness are inscribed in the cone, one on top of the other, as suggested in Fig. 25. Of course, the sum of the

FIG. 25

volumes of these discs will be less than the volume of the cone. However, as the number of discs increases and approaches infinity, and correspondingly the thickness of each individual disc shrinks and approaches zero, then clearly the sum of the volumes of all the discs will approach the exact volume of the cone.

To carry out the details of this idea, let $n$ be a given large positive integer. If there are $n - 1$ inscribed discs, then each disc has thickness $\frac{h}{n}$, as shown in Fig. 26. If $y$ is the distance down from the vertex of the cone to the upper surface of the

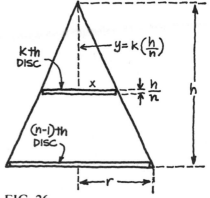

$y = k\left(\frac{h}{n}\right)$

$kth$
$DISC$

$x$

$\frac{h}{n}$

$h$

$(n-1)th$
$DISC$

$r$

FIG. 26

$k$th disc, where $k$ is any one of the numbers 1, 2, $\ldots, n-1$, then $y = k\left(\dfrac{h}{n}\right)$. If $x$ is the radius of the $k$th disc, then by similar triangles we have

$$\frac{x}{y} = \frac{r}{h}, \quad \text{so} \quad x = \left(\frac{r}{h}\right)y = \left(\frac{r}{h}\right)k\left(\frac{h}{n}\right) = \frac{kr}{n}.$$

The volume of the $k$th disc is therefore

$$\pi x^2 \left(\frac{h}{n}\right) = \pi \frac{k^2 r^2}{n^2}\left(\frac{h}{n}\right) = \frac{\pi r^2 h}{n^3} k^2,$$

and the sum of these quantities for $k = 1, 2, \ldots,$ $n-1$ is the total volume $v$ of all the inscribed discs:

$$v = \frac{\pi r^2 h}{n^3}[1^2 + 2^2 + \cdots + (n-1)^2].$$

By using formula (2) with $n$ replaced by $n-1$, we can write this as

$$v = \frac{\pi r^2 h}{n^3} \cdot \frac{(n-1)n(2n-1)}{6}$$

$$= \frac{\pi r^2 h}{6}\left(1 - \frac{1}{n}\right)\left(2 - \frac{1}{n}\right).$$

As $n$ approaches infinity, it is easy to see that this approaches $\frac{1}{3}\pi r^2 h$, the exact volume of the cone.

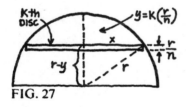

FIG. 27

To establish the formula for the volume of the sphere in Fig. 24, we apply a similar method to the upper hemisphere, which is shown in profile in Fig. 27. This time, the radius $x$ of the $k$th disc is found by using the Pythagorean theorem,

$$x^2 = r^2 - (r-y)^2 = 2ry - y^2$$

$$= 2rk\left(\frac{r}{n}\right) - k^2\left(\frac{r}{n}\right)^2.$$

The volume of the $k$th disc is therefore

$$\pi x^2 \left(\frac{r}{n}\right) = \pi\left[2rk\left(\frac{r}{n}\right) - k^2\left(\frac{r}{n}\right)^2\right]\left(\frac{r}{n}\right)$$

$$= \frac{2\pi r^3}{n^2}k - \frac{\pi r^3}{n^3}k^2,$$

and the sum of these quantities for $k = 1, 2, \ldots,$ $n-1$ is the total volume $v$ of all the inscribed discs:

$$v = \frac{2\pi r^3}{n^2}[1 + 2 + \cdots + (n-1)]$$

$$- \frac{\pi r^3}{n^3}[1^2 + 2^2 + \cdots + (n-1)^2].$$

By using formulas (1) and (2) with $n$ replaced by $n-1$, we can write this as

ALGEBRA

$$v = \frac{2\pi r^3}{n^2} \cdot \frac{(n-1)n}{2} - \frac{\pi r^3}{n^3} \cdot \frac{(n-1)n(2n-1)}{6}$$

$$= \pi r^3\left(1 - \frac{1}{n}\right) - \frac{\pi r^3}{6}\left(1 - \frac{1}{n}\right)\left(2 - \frac{1}{n}\right).$$

As $n$ approaches infinity, it is easy to see that this approaches

$$\pi r^3 - \frac{1}{3}\pi r^3 = \frac{2}{3}\pi r^3,$$

so the volume of the complete sphere is $\frac{4}{3}\pi r^3$.

# ANSWERS
## SECTION 1

1. $-\frac{\pi}{3}$ and $\frac{\sqrt{5}}{2}$ are irrational; $-\frac{2}{3}$ and $\frac{5}{1234}$ are rationals; $-10$ and $-\sqrt{4} = -2$ are integers; $\sqrt{9}$ $= 3$ and $\frac{51}{3} = 17$ are positive integers.

2. (a) $2a$; (b) 0; (c) $3b - 4a$.
3. (a) 3401; (b) 9180; (c) 3322.
4. (a) $6(2x - 3y + 5)$; (b) $4x^2(2 - 3xy - 7x^2z)$;

(c) $3abc(3 + abc)$.   5. (a) $\frac{a^2 - b^2}{ab}$; (b) $\frac{2}{x-2}$;

(c) $\frac{x-1}{x}$; (d) $\frac{x^2 + y^2}{x^2y^2}$; (e) $\frac{16a^2 + b^2}{4ab}$.

6. (a) $\frac{5}{a^3}$; (b) $\frac{1}{125a^3}$; (c) 9; (d) 1.

7. (a) $a^n b^{4n}$; (b) $12a^2b$; (c) $x^{10}y^{10}$; (d) $b^2 + a^2$;

(e) $\frac{(x+y)^2}{xy}$; (f) 1.   8. (a) 7; (b) 12; (c) 5;

(d) 10; (e) 3; (f) 3; (g) 2; (h) .8; (i) .3; (j) $\frac{4}{11}$;

(k) $\frac{3}{4}$; (l) $-\frac{1}{3}$; (m) $\frac{4}{5}$; (n) $-10$; (o) $5\sqrt{5}$; (p) 25;

(q) 5; (r) $3\sqrt{2}$; (s) $2\sqrt{3}$; (t) $3\sqrt{2}$; (u) $2\sqrt{3}$;

(v) $8\sqrt[3]{2}$; (w) $\sqrt{2a}$; (x) $ab^2$; (y) $a\sqrt[4]{a}$; (z) $\frac{1}{2}$.

9. (a) $5\sqrt{6}$; (b) $5 + 2\sqrt{6}$; (c) $\sqrt{7} - \sqrt{5}$.

10. (a) 6; (b) 2; (c) 16; (d) 216; (e) 36; (f) $\frac{1}{4}$;

(g) $\frac{1}{27}$; (h) $\frac{1}{4}$; (i) 1000; (j) 27; (k) 10.

11. (a) $\frac{5a^3}{b}$; (b) $16a^2b$; (c) $ab$; (d) $\frac{a^2}{6}$; (e) $5\sqrt[3]{a}$;

(f) $a - b$; (g) $a^3$; (h) $\dfrac{3b^2c^2}{4a^2}$.

**12.** (a) $x^7 + 2x^6 - 8x^5 - 2x^4 + x^3 + 2x^2 + x - 8$;
(b) $3x^5 - 2x^3 - 11x$.   **13.** (a) $6x^5 + 5x^4 - 24x^3 - 39x^2 + 8x + 36$; (b) $2x^7 - 8x^6 + 16x^4 - 8x^3 + 6x^2 - 24x + 12$; (c) $x^3 - 1$; (d) $x^4 - 1$; (e) $x^5 - 1$.

**14.** (a) $(x - 3)(x + 2)$; (b) $(x + 4)(x + 5)$;
(c) $(x + 2)(x + 10)$; (d) $(x - 2)^2$; (e) $(x + 4)^2$;
(f) $x(x + 6)^2$; (g) $(x^2 + 4)(x + 2)(x - 2)$;
(h) $(x + 15)(x - 2)$; (i) $(x + 7)(x - 5)$;
(j) $(x - 6)(x - 7)$; (k) $x(x - 4)(x + 1)$;
(l) $(2x + 4)(2x - 3)$; (m) $(2x - 4)(5x + 2)$.

**15.** (a) $(x - 3)(x^2 + 3x + 9)$; (b) $(2x - 5)$ $\times (4x^2 + 10x + 25)$.

**16.** (a) $(x + 4)(x^2 - 4x + 16)$; (b) $(3x + 2)$ $\times (9x^2 - 6x + 4)$.   **17.** (a) $4, -7$; (b) $11, -3$;
(c) $\dfrac{5}{2}, -3$; (d) $\dfrac{7}{3}, -\dfrac{3}{2}$.   **18.** (a) $\dfrac{9 \pm \sqrt{21}}{10}$;
(b) $\dfrac{-7 \pm \sqrt{13}}{6}$; (c) $\dfrac{3 \pm 2\sqrt{-2}}{17}$; (d) $\dfrac{-1 \pm \sqrt{-3}}{2}$.

**19.** (a) $3 < 11$; (b) $5 > 2$; (c) $-4 < 3$;
(d) $-6 < -2$; (e) $-2 > -3$; (f) $\pi < 2\sqrt{3}$.
**20.** (a) $x < -6$; (b) $x > 3$.   **21.** (a) $x = \pm 2$;
(b) $x = \pm 3$; (c) $x = \pm 6$; (d) $x = -1, 5$;
(e) $x = -4, -2$.   **22.** $x > 1$ or $x < 0$.
**23.** $x > 3$ or $x < -5$.

# SECTION 2

**1.** $-3, 1, 5, 9$.   **2.** $-4, -\dfrac{1}{2}, -\dfrac{20}{7}$.
**3.** $x^9 - 3x^6 + 5x^3 - 1$.   **4.** $x$.   **5.** (a) $A = x^2$;
(b) $A = \dfrac{p^2}{16}$   **6.** $A = \dfrac{c^2}{4\pi}$.   **7.** $h = \dfrac{1}{2}\sqrt{3}x$.
**8.** $d = 5t$.   **10.** (a) $3, 4, 5$; (b) $12, 5, 13$.
**11.** (a) $12$; (b) $35$.   **12.** (a) $(3, 4)$; (b) $(4, 5)$.
**13.**

(a)                (b)              (c)

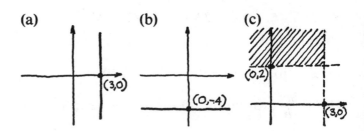

(d)　　　　(e)　　　　(f)

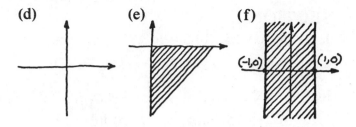

(-1,0)　　　　(1,0)

14. (a) 2; (b) 0; (c) −1.　15. (a) $\dfrac{y+2}{x+1}=1$ or

$y=x-1$; (b) $y=3x+4$; (c) $x=4$.

17. (a) $\dfrac{y+3}{x+2}=-\dfrac{1}{2}$ or $x+2y=-8$; (b) $\dfrac{y+3}{x+2}=2$

or $y=2x+1$.　18. (a) $x^2+y^2=4$; (b) $(x+2)^2$

$+y^2=49$; (c) $(x-3)^2+(y-6)^2=\dfrac{1}{4}$;

(d) $(x-1)^2+(y-2)^2=25$.　19. (a) $(-3,6)$, 3;

(b) $(4,0)$, 2; (c) $(0,-2)$, 1; (d) $(-3,-1)$, 2;

(e) $(8,-7)$, 4.　20. $(-2,2)$ and $(5,3)$.

21. (a) $\left(0,\dfrac{1}{8}\right)$, $y=-\dfrac{1}{8}$; (b) $(0,2)$, $y=-2$;

(c) $\left(0,-\dfrac{1}{20}\right)$, $y=\dfrac{1}{20}$; (d) $(0,-3)$, $y=3$.

22. (a) $x^2=12y$, (b) $x^2=64y$; (c) $x^2=-4y$;

(d) $5x^2=-2y$.　23. (a) $(2,-3)$, $\left(2,-2\dfrac{3}{4}\right)$, up;

(b) $(3,-25)$, $\left(3,-24\dfrac{7}{8}\right)$, up; (c) $(-2,9)$,

$\left(-2,8\dfrac{3}{4}\right)$, down; (d) $(-2,6)$, $\left(-2,5\dfrac{1}{2}\right)$, down.

24. (a) imaginary; (b) real and distinct; (c) real

and distinct; (d) real and equal.　25. (a) sum $=$

$\dfrac{7}{4}$, product $=-\dfrac{13}{4}$; (b) we don't know what it

means to add or multiply imaginary numbers;

(c) sum $=-\dfrac{1}{2}$, product $=-1$.　26. (a) $x^2-5x-$

$24=0$; (b) $x^2-4x-1=0$; (c) $15x^2-x-6=0$.

27.

(a)　　　　(b)　　　　(c)　　　　(d)

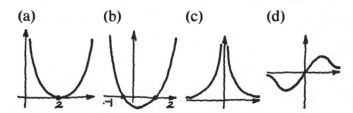

## SECTION 3

**1.** (a) $2 = \log_5 25$; (b) $5 = \log_2 32$; (c) $-2 = \log_5 \frac{1}{25}$; (d) $6 = \log_2 64$; (e) $3 = \log_3 27$; (f) $.5 = \log_{81} 9$; (g) $0 = \log_7 1$; (h) $-1 = \log_{10} \frac{1}{10}$; (i) $\frac{4}{5} = \log_{32} 16$; (j) $.75 = \log_{16} 8$.   **2.** (a) $10^1 = 10$; (b) $4^{3/2} = 8$; (c) $2^3 = 8$; (d) $5^{-3} = \frac{1}{125}$; (e) $9^2 = 81$; (f) $10^0 = 1$; (g) $10^{-2} = .01$; (h) $7^3 = 343$; (i) $5^3 = 125$; (j) $10^{-1} = .1$.   **3.** (a) 1; (b) 4; (c) 4; (d) $\frac{3}{2}$; (e) $-3$; (f) $\frac{2}{3}$; (g) 0; (h) $\frac{7}{9}$; (i) $\frac{7}{6}$.

**4.** (a) 2187; (b) 243; (c) 256; (d) 16.   **5.** (a) 243; (b) $\frac{1}{81}$; (c) 7; (d) 36.   **6.** (a) $Q(x) = x - 2$, $R(x) = 0$; (b) $Q(x) = x + 3$, $R(x) = 0$; (c) $Q(x) = x^2 - x - 6$, $R(x) = 0$; (d) $Q(x) = 3x^2 - 2x + 1$, $R(x) = -x + 7$; (e) $Q(x) = x^3 - 2x^2 + 5$, $R(x) = 0$.   **8.** $x^{n-1} + x^{n-2} + x^{n-3} + \cdots + x + 1$.   **9.** (a) $(x - 1)(x - 2)(x - 3)$; (b) $(x - 1)(x^2 + 1)$; (c) $(x + 2)(x + 6)(x - 3)$; (d) $(x - 3)(x + 1)^2$.   **10.** (a) $-2$; (b) 47.   **11.** (a) $-83$; (b) 108.   **12.** (a) $x = -4$, $y = 6$; (b) $x = -3$, $y = 2$, $z = 4$.   **13.** $5 \cdot 4^7 = 81,920$.

**14.** 325 inches.   **15.** (a) 2; (b) $\frac{8}{3}$; (c) 27; (d) $\frac{9}{2}$; (e) $\frac{3\sqrt{3}(\sqrt{3} + 1)}{2}$; (f) $2\sqrt{6}(\sqrt{2} + 1)$.

**16.** (a) $\frac{7}{9}$; (b) $\frac{34}{99}$; (c) $\frac{838}{225}$.   **17.** 405 inches.

**18.** 24 inches.   **20.** $n^2$.   **21.** (a) 504; (b) 1,235,520; (c) 816; (d) 58,905.

**22.** 40,320; 336.   **23.** 2184.   **24.** 84.   **25.** 5005.

**26.** (a) $a^9 + 9a^8 b + 36a^7 b^2 + 84a^6 b^3 + 126a^5 b^4 + 126a^4 b^5 + 84a^3 b^6 + 36a^2 b^7 + 9ab^8 + b^9$; (b) $64a^6 + 192a^5 b + 240a^4 b^2 + 160a^3 b^3 + 60a^2 b^4 + 12ab^5 + b^6$; (c) $32a^5 - 240a^4 b + 720a^3 b^2 - 1080a^2 b^3 + 810ab^4 - 243b^5$.

# CHAPTER 3
# TRIGONOMETRY

"Any two philosophers can tell each other all they know in two hours."
                        — Oliver Wendell Holmes, Jr.

# INTRODUCTION

I have been teaching trigonometry for more than 30 years, as a brief but essential interlude in calculus courses. In every such course there comes a time when it is necessary for the students to have a reasonable understanding of the trigonometric functions and their properties. And in an effort to make certain that everyone in my classes is properly prepared, I routinely cover everything that matters in trigonometry—from the beginning, with proofs—in a single 50-minute lecture.

Under these circumstances, it was only natural for me to ask myself why this subject consumes an entire semester in most high school curricula. I have also asked myself why some trigonometry textbooks are nearly 500 pages long. When I looked into some of these books recently I found the answers to both questions: hundreds of pages of unnecessary padding, consisting mostly of obscure formalities and irrelevant digressions. This padding is like smog, or dust in the eyes—it makes it almost impossible for even bright students to gain any clear view of what the subject is about or what it is for.

This chapter is a slightly expanded version of my standard 50-minute lecture. I have written it out for two reasons. First, I would like to give my students—and perhaps other calculus students as well—something definite they can refer to, in order to fix the ideas of trigonometry more firmly in mind. And second, I hope that high school teachers will find it a worthwhile tool for their students to use in achieving a solid grasp of the main points of the subject. If I can cover this material in 50 minutes of unhurried speaking, a determined student should be able to master it in a few hours. At least, the student will have no difficulty here in separating the wheat from the chaff—for there is no chaff.

# 1. MISCONCEPTIONS OF THE NATURE OF THE SUBJECT

Most trigonometry textbooks have been written by people who appear to believe that the importance of the subject lies in its applications to surveying and navigation. Even though very few people become surveyors or navigators, the students who study these books are expected to undertake many lengthy calculations about the heights of flagpoles, the widths of rivers and the positions of ships at sea.

The truth is that the primary importance of trigonometry lies in a completely different direction — in the mathematical description of vibrations, rotations, and periodic phenomena of all kinds, including light, sound, alternating currents, and the orbits of the planets around the sun. What matters most in the subject is not making computations about triangles, but grasping the *trigonometric functions* as indispensable tools in science, engineering and higher mathematics. These functions and their properties are the sole subject matter of this chapter.

# 2. A FEW NECESSARY FACTS FROM GEOMETRY AND ALGEBRA

We will need to have an easy familiarity with the following basic facts from elementary geometry:

(a) The sum of the angles in any triangle equals 180 degrees (180°). This is illustrated on the left in Fig. 1, and is proved on the right with the aid of the dotted auxiliary lines.

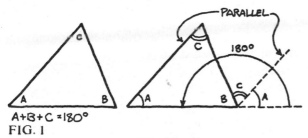

$A+B+C = 180°$
FIG. 1

(b) Roughly speaking, two triangles are *similar* if they have the same shape but different sizes, that is, if one is a magnified version of the other.

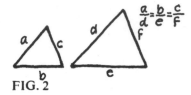

FIG. 2

The precise meaning of similarity for triangles is that their corresponding angles must be equal; and this implies that the ratios of their corresponding sides must also be equal, as shown in Fig. 2. The first of these equations $\left(\dfrac{a}{d} = \dfrac{b}{e}\right)$ can be written in the equivalent form

$$\frac{a}{b} = \frac{d}{e}.$$

In words: if two triangles are similar then the ratio of any two sides of one triangle equals the ratio of the corresponding sides of the other. By part (a) above, two triangles will necessarily be similar if two pairs of their corresponding angles are equal.

(c) The *Pythagorean theorem* states that in any right triangle (one angle is a right angle, a 90° angle) the square of the hypotenuse equals the sum of the squares of the legs. This is illustrated on the left in Fig. 3, and is proved on the right.

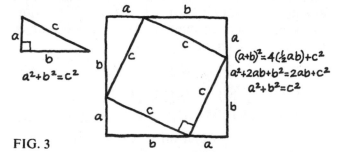

FIG. 3

(The proof is carried out by inserting four replicas of the triangle in the corners of a square of side $a + b$. The first equation on the right states that the area of the large square equals the area of the four triangles plus the area of the small inner square.)

(d) There are two special triangles whose properties will play a role in our work: the isosceles right triangle on the left in Fig. 4, and the 30°-60°

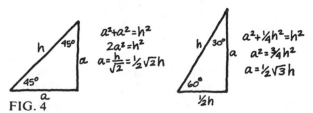

FIG. 4

94

right triangle on the right. In the indicated calculations we find the legs in terms of the given hypotenuse $h$, using the Pythagorean theorem and the fact that in a 30°-60° right triangle the side opposite the 30° angle is half the hypotenuse. Since $\sqrt{2}$ and $\sqrt{3}$ are approximately 1.414 and 1.732, the heights of the two triangles are approximately $.707h$ and $.866h$.*

(e) If $(x_1, y_1)$ and $(x_2, y_2)$ are two given points in the $xy$-plane of analytic geometry, as shown in Fig. 5, then the numbers $|x_1 - x_2|$ and $|y_1 - y_2|$ are the legs of the indicated right triangle.† If $d$ is the distance between the points, then the Pythagorean theorem tells us that

$$d^2 = |x_1 - x_2|^2 + |y_1 - y_2|^2$$
$$= (x_1 - x_2)^2 + (y_1 - y_2)^2.$$

This yields the *distance formula:*
$$d = \sqrt{(x_1 - x_2)^2 + (y_1 - y_2)^2}.$$

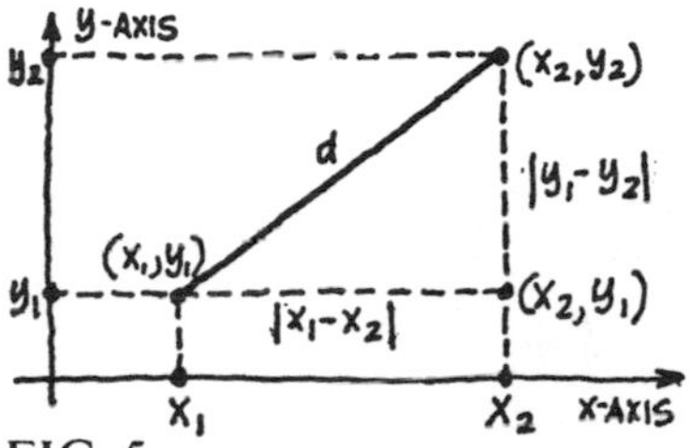

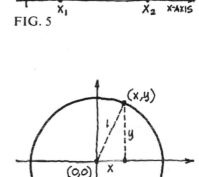

FIG. 5

(f) The set of all points $(x, y)$ whose distance from the origin $(0, 0)$ is equal to 1 is a circle of radius 1 called the *unit circle*. It is clear from Fig. 6 and the Pythagorean theorem that

$$x^2 + y^2 = 1$$

is the equation of the unit circle.

(g) The primary purpose of this chapter is to give a brief but clear account of the trigonometric functions; and to understand such an account, it is first necessary to understand what a function is.

FIG. 6

If $x$ and $y$ are two variables that are related in such a way that whenever a numerical value is assigned to $x$ there is determined one and only one corresponding numerical value for $y$, then $y$ is called a *function* of $x$ and this is symbolized by writing $y = f(x)$. [The letter $f$ symbolizes the operation, or rule of correspondence, that yields

---

*The process by which $\dfrac{h}{\sqrt{2}}$ is written as $\dfrac{h}{\sqrt{2}} = \dfrac{\sqrt{2}h}{\sqrt{2}\sqrt{2}}$
$= \dfrac{1}{2}\sqrt{2}h$ is called "rationalizing the denominator." Its purpose is to make it easier for us to compute decimals (it is clearly easier to divide 1.414 by 2 than to divide 1 by 1.414).
†Recall that the *absolute value* of any number $a$, denoted by $|a|$, is defined by $|a| = \begin{cases} a \text{ if } a \geq 0, \\ -a \text{ if } a < 0. \end{cases}$ Thus, $|2| = 2$, $|0| = 0$, and $|-3| = -(-3) = 3$.

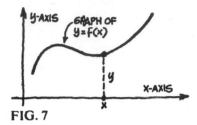

FIG. 7

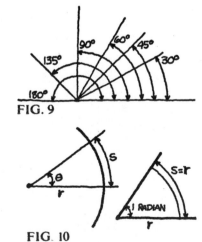

FIG. 9

FIG. 10

$y$ when applied to $x$.] The most convenient way of visualizing a function is by means of its *graph*, as shown in Fig. 7. The *independent variable x* can be thought of as a point moving along the *x*-axis from left to right; each $x$ determines a value of the *dependent variable y*, which is the height of the point $(x, y)$ above the *x*-axis; and the graph of the function is simply the path of the point $(x, y)$ as it moves across the plane and varies in height according to the nature of the particular function under consideration.

There are many types of functions. The most familiar are those defined by simple algebraic formulas. We mention three examples:

$$y = 2x - 1, \qquad y = x^2, \qquad \text{and} \qquad y = \sqrt{x}.$$

The graphs of these functions are shown in Fig. 8.

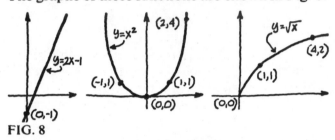

FIG. 8

The best method for graphing such a function is to plot a few carefully selected points and then sketch in the qualitative features of the graph as these are revealed by an examination of the form of the function.

## 3. RADIAN MEASURE FOR ANGLES

Angles can be measured in many different ways. One familiar system uses the *degree* as the basic unit, where one degree (1°) is one-ninetieth of a right angle. We have assumed that the reader is already acquainted with the use of degrees (Fig. 9).

However, the *radian* is the natural unit in all of higher mathematics, and in most of science and engineering. To understand what the radian system is, denote the angle under discussion by the Greek letter $\theta$ (theta) and place it at the center of a circle of radius $r$ (Fig. 10, left). If $s$ is the length of the arc subtended by the sides of the

96

angle, measured in the same linear units as the radius $r$, then the number of radians in $\theta$ is defined to be the number $\frac{s}{r}$. We can therefore write

$$\theta = \frac{s}{r}$$

if the angle $\theta$ is understood to be measured in radians. It should be noticed that an angle of 1 radian (Fig. 10, right) cuts off an arc exactly equal in length to the radius. As we explain below, 1 radian equals approximately 57.3 degrees. To find the relation between the number of degrees and the number of radians in an angle, we recall that the circumference of a circle is given by the formula $c = 2\pi r$, where $\pi$ is approximately 3.1416.* The length of a semicircular arc is therefore $\pi r$

(Fig. 11); and since $\frac{\pi r}{r} = \pi$, we have the basic relation

$180° = \pi$ radians.

On dividing this by 180 we get

$1° = \frac{\pi}{180}$ radians $= .017$ radians, approximately;

and dividing by $\pi$ gives

1 radian $= \left(\frac{180}{\pi}\right)^° = 57.3°$, approximately.

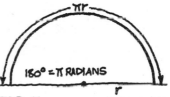

FIG. 11

By appropriate divisions many other equivalences can be found; for example,

$90° = \frac{\pi}{2}$ radians   and   $60° = \frac{\pi}{3}$ radians.

It is a universal convention that when an angle is measured in radians, no unit of angular measure is mentioned. Thus, an angle $\frac{\pi}{4}$ means an angle

of $\frac{\pi}{4}$ radians, and an angle 2 means an angle of 2 radians. We list the radian measures of a few angles that are often encountered:

$30° = \frac{\pi}{6}$,     $45° = \frac{\pi}{4}$,     $60° = \frac{\pi}{3}$,     $90° = \frac{\pi}{2}$,

---

*This formula is true because it is simply a rewritten version of the definition of $\pi$, namely,

$$\pi = \frac{\text{circumference of circle}}{\text{diameter of circle}} = \frac{c}{2r}.$$

$$120° = \frac{2\pi}{3}, \qquad 135° = \frac{3\pi}{4}, \qquad 180° = \pi,$$

$$270° = \frac{3\pi}{2}, \qquad 360° = 2\pi.$$

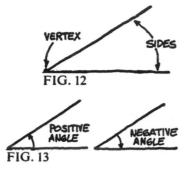

FIG. 12

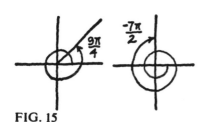

FIG. 13

So far, we have considered an angle to be a geometric figure consisting of two rays (half-lines), called the sides, which have a common endpoint called the vertex (Fig. 12). However, the angles we consider are usually *directed*. This means that one side is designated as the *initial side,* the other side as the *terminal side,* and the angle is thought of as generated by rotating the initial side to the terminal side. It is customary to call a *counterclockwise* rotation positive and a *clockwise* rotation negative. With this agreement, we call the measure of an angle *positive* or *negative* according to the direction in which its initial side is rotated to its terminal side (Fig. 13). The angles in Fig. 14 are placed in *standard position* in the *xy*-plane,

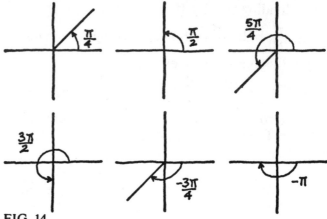

FIG. 14

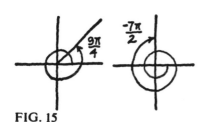

FIG. 15

which means that each has its vertex at the origin and its initial side lying along the positive *x*-axis. It is possible that there might be more than one complete rotation involved in sending the initial side to the terminal side. Two such angles are shown in Fig. 15.

## 4. THE TRIGONOMETRIC FUNCTIONS

In older textbooks the sine, cosine and tangent of an acute angle $\theta$ were often defined by means of a right triangle, as follows. Draw a right tri-

## TRIGONOMETRY

angle with $\theta$ as one of its acute angles (Fig. 16) and define sin $\theta$, cos $\theta$, tan $\theta$ to be the following ratios of the sides of this triangle:

$$\sin \theta = \frac{\text{opposite side}}{\text{hypotenuse}} = \frac{a}{h},$$

$$\cos \theta = \frac{\text{adjacent side}}{\text{hypotenuse}} = \frac{b}{h},$$

$$\tan \theta = \frac{\text{opposite side}}{\text{adjacent side}} = \frac{a}{b}.$$

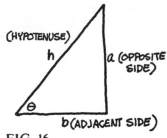

FIG. 16

The first two of these definitions, when written in the form

$$a = h \sin \theta \quad \text{and} \quad b = h \cos \theta,$$

have many uses in physics and geometry. Nevertheless, this right triangle approach to trigonometry has the fundamental defect that the angle $\theta$ must be positive and less than $\frac{\pi}{2}$ for these definitions to make sense—whereas all of the more advanced applications to science, engineering and mathematics require that $\theta$ be unrestricted. We remove this crippling limitation on $\theta$ by adopting the following more satisfactory definitions.

Consider the unit circle $x^2 + y^2 = 1$ in the $xy$-plane (Fig. 17). Let $\theta$ be any number whatever, positive or negative, large or small. Generate an angle $\theta$ by a suitable rotation of the radius lying along the positive $x$-axis—counterclockwise through $\theta$ radians if $\theta$ is positive, and clockwise through $-\theta$ radians if $\theta$ is negative. With these agreements, the endpoint $(x, y)$ of the terminal side of the angle is uniquely determined when the number $\theta$ is given. This endpoint can lie in any one of the four quadrants, and its coordinates $x$ and $y$ can be positive or negative in any combination. Apart from their algebraic signs, $x$ and $y$ can be thought of as the base and height of the right triangle whose hypotenuse is the terminal side of the angle $\theta$. The six trigonometric functions (the three already mentioned, together with three more—the cotangent, secant, cosecant) are now defined as follows:

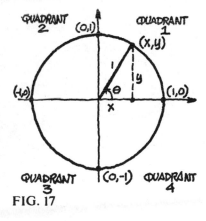

FIG. 17

$$\sin \theta = y, \quad \tan \theta = \frac{y}{x}, \quad \sec \theta = \frac{1}{x},$$

$$\cos \theta = x, \quad \cot \theta = \frac{x}{y}, \quad \csc \theta = \frac{1}{y}.$$

99

We have drawn the angle $\theta$ in Fig. 17 to be equal to the angle $\theta$ in Fig. 16, so that the two triangles will be similar. Our reason for doing this is to point out that the unit circle definitions for sin $\theta$, cos $\theta$, tan $\theta$ agree with the right triangle definitions when $\theta$ is a positive acute angle. To see this, notice in Fig. 17 that

$$\sin \theta = y = \frac{y}{1} = \frac{\text{opposite side}}{\text{hypotenuse}}, \text{ etc.}$$

We therefore have no conflict between the right triangle definitions and the unit circle definitions, and the latter are much broader in scope.

The first two of our six functions are the basic ones, and if necessary each of the others can be expressed in terms of these two. These relations (and one more) are immediate consequences of the above definitions:

$$\tan \theta = \frac{\sin \theta}{\cos \theta}, \tag{1}$$

$$\cot \theta = \frac{\cos \theta}{\sin \theta}, \tag{2}$$

$$\sec \theta = \frac{1}{\cos \theta}, \tag{3}$$

$$\csc \theta = \frac{1}{\sin \theta}, \tag{4}$$

$$\cot \theta = \frac{1}{\tan \theta}. \tag{5}$$

These are the first of the 21 fundamental identities that express the interrelations and properties of the trigonometric functions and constitute the heart of the subject. They fall into several natural groups, and are therefore easier to remember than might be expected. We emphasize these groups by enclosing them in boxes.

Our next identities state the effect of replacing $\theta$ by $-\theta$. As angles, $\theta$ and $-\theta$ are generated by rotations of equal magnitude but opposite directions, so the endpoints of their terminal sides lie on the same vertical line (Fig. 18). This remark proves the first two of the following identities:

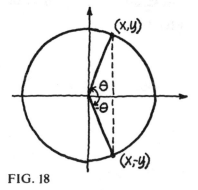

FIG. 18

$$\boxed{\begin{aligned} \sin{(-\theta)} &= -\sin{\theta}, \\ \cos{(-\theta)} &= \cos{\theta}, \\ \tan{(-\theta)} &= -\tan{\theta}. \end{aligned}}$$

        (6)
(7)
(8)

The third follows at once from (1) combined with (6) and (7).*

 Before writing down our third group of identities, we must explain that the notations $\sin^2\theta$ and $\cos^2\theta$ mean $(\sin\theta)^2$ and $(\cos\theta)^2$. If the equation $x^2 + y^2 = 1$ of the unit circle is written in the form $y^2 + x^2 = 1$, then this translates directly into $\sin^2\theta + \cos^2\theta = 1$.

If we first divide this through by $\cos^2\theta$, and then by $\sin^2\theta$, we obtain

$$\left(\frac{\sin\theta}{\cos\theta}\right)^2 + 1 = \left(\frac{1}{\cos\theta}\right)^2$$

and

$$1 + \left(\frac{\cos\theta}{\sin\theta}\right)^2 = \left(\frac{1}{\sin\theta}\right)^2.$$

These remarks yield the identities

$$\boxed{\begin{aligned} \sin^2\theta + \cos^2\theta &= 1, \\ \tan^2\theta + 1 &= \sec^2\theta, \\ 1 + \cot^2\theta &= \csc^2\theta, \end{aligned}}$$

        (9)
(10)
(11)

which are clearly equivalent versions of one another.

# 5. THE VALUES OF sin $\theta$, cos $\theta$ AND tan $\theta$ FOR CERTAIN SPECIAL ANGLES

 If we keep firmly in mind the definitions of $\sin\theta$, $\cos\theta$ and $\tan\theta$, then there are several 1st quadrant angles (Fig. 19) for which the values of these functions are easy to find by means of the ideas in part (d) of Section 2. All that is necessary is to look with understanding at the three parts of Fig. 19:

---

*It is clear that there are similar identities for the cotangent, secant and cosecant. However, these are of little importance, and in keeping with our policy of rigorously excluding all nonessentials, we disregard them.

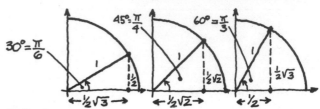

FIG. 19

$$\sin\frac{\pi}{6}=\frac{1}{2}, \qquad \cos\frac{\pi}{6}=\frac{1}{2}\sqrt{3}, \qquad \tan\frac{\pi}{6}=\frac{\frac{1}{2}}{\frac{1}{2}\sqrt{3}}=\frac{1}{\sqrt{3}}=\frac{1}{3}\sqrt{3};$$

$$\sin\frac{\pi}{4}=\frac{1}{2}\sqrt{2}, \qquad \cos\frac{\pi}{4}=\frac{1}{2}\sqrt{2}, \qquad \tan\frac{\pi}{4}=\frac{\frac{1}{2}\sqrt{2}}{\frac{1}{2}\sqrt{2}}=1;$$

$$\sin\frac{\pi}{3}=\frac{1}{2}\sqrt{3}, \qquad \cos\frac{\pi}{3}=\frac{1}{2}, \qquad \tan\frac{\pi}{3}=\frac{\frac{1}{2}\sqrt{3}}{\frac{1}{2}}=\sqrt{3}.$$

Also, of course, sin 0 = 0, cos 0 = 1, and tan 0 = 0; and $\sin\frac{\pi}{2}=1$, $\cos\frac{\pi}{2}=0$, and $\tan\frac{\pi}{2}=\frac{1}{0}$ is undefined. These facts and a few others are organized in the following table. This information is best learned, not by memorizing it, but by knowing the definitions and visualizing the appropriate pictures.

| $\theta$ | 0 | $\frac{\pi}{6}$ | $\frac{\pi}{4}$ | $\frac{\pi}{3}$ | $\frac{\pi}{2}$ | $\frac{2\pi}{3}$ | $\frac{3\pi}{4}$ | $\frac{5\pi}{6}$ | $\pi$ | $\frac{3\pi}{2}$ | $2\pi$ |
|---|---|---|---|---|---|---|---|---|---|---|---|
| $\sin\theta$ | 0 | $\frac{1}{2}$ | $\frac{1}{2}\sqrt{2}$ | $\frac{1}{2}\sqrt{3}$ | 1 | $\frac{1}{2}\sqrt{3}$ | $\frac{1}{2}\sqrt{2}$ | $\frac{1}{2}$ | 0 | −1 | 0 |
| $\cos\theta$ | 1 | $\frac{1}{2}\sqrt{3}$ | $\frac{1}{2}\sqrt{2}$ | $\frac{1}{2}$ | 0 | $-\frac{1}{2}$ | $-\frac{1}{2}\sqrt{2}$ | $-\frac{1}{2}\sqrt{3}$ | −1 | 0 | 1 |
| $\tan\theta$ | 0 | $\frac{1}{3}\sqrt{3}$ | 1 | $\sqrt{3}$ | * | $-\sqrt{3}$ | −1 | $-\frac{1}{3}\sqrt{3}$ | 0 | * | 0 |

If we know the values of sin $\theta$, cos $\theta$ and tan $\theta$ for 1st quadrant angles, we can easily find their values for angles in any quadrant. There are two key ideas. First, we must understand the way the algebraic signs of these functions change from one quadrant to another. The facts are obvious from Fig. 17, and are stated in the following table.

**TRIGONOMETRY**

| quadrant | 1 | 2 | 3 | 4 |
|---|---|---|---|---|
| sin θ | + | + | − | − |
| cos θ | + | − | − | + |
| tan θ | + | − | + | − |

Second, we must determine the positive acute angle $\theta_0$ formed by the terminal side of $\theta$ and the nearest half of the x-axis (Fig. 20). The desired

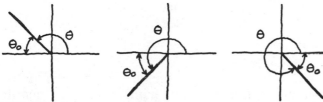

FIG. 20

value will then be one of the numbers $\pm\sin\theta_0$, $\pm\cos\theta_0$, $\pm\tan\theta_0$, with the sign determined by the quadrant of $\theta$. For example, if $\theta=\frac{11\pi}{6}$ $(=330°)$ then $\theta_0=\frac{\pi}{6}$; and since $\theta$ is a 4th quadrant angle,

$$\sin\theta=-\sin\theta_0=-\sin\frac{\pi}{6}=-\frac{1}{2}.$$

# 6. THE GRAPHS OF sin θ, cos θ AND tan θ

The graph of sin θ is easy to draw by looking at Fig. 17 and following the way y varies as θ increases from 0 to 2π, that is, as the radius swings around through one complete counterclockwise rotation. This gives one complete cycle of sin θ, as shown on the left in Fig. 21. The complete graph (on the right in Fig. 21) consists of

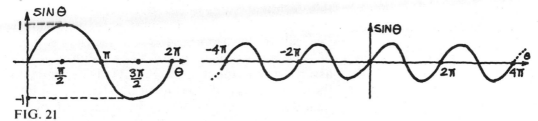

FIG. 21

infinitely many repetitions of this cycle, to the right and to the left. The graph of cos θ can be sketched in essentially the same way (Fig. 22).

103

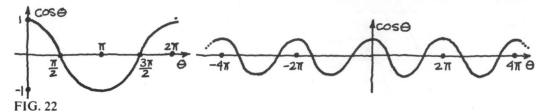

FIG. 22

It is obvious from Fig. 17 that the angles $\theta$ and $\theta + 2\pi$ have the same terminal sides. We therefore have

$$\sin (\theta + 2\pi) = \sin \theta \quad \text{and} \quad \cos (\theta + 2\pi) = \cos \theta,$$

and for this reason we say that each function is *periodic* with *period* $2\pi$.

If $b$ is a positive constant, then the function $\sin b\theta$ goes through one complete cycle when the quantity $b\theta$ increases from 0 to $2\pi$, that is, when the variable $\theta$ increases from 0 to $\dfrac{2\pi}{b}$. This function is therefore periodic with period $\dfrac{2\pi}{b}$. If we now multiply $\sin b\theta$ by a positive constant $a$, the effect is to magnify the graph by a factor of $a$ in the vertical direction. The peaks of $a \sin b\theta$ are therefore $a$ units above the $\theta$-axis, and for this reason $a$ is called the *amplitude* of the function $a \sin b\theta$.

The graph of $\tan \theta$ is easy to sketch with the aid of the following geometric interpretation. Fig. 23 is Fig. 17 with the addition of a vertical line drawn through the point $(1, 0)$, tangent to the circle at this point. We introduce a coordinate system on this line by specifying that the coordinate of a point on the line is simply its $y$-coordinate as a point in the plane. Now, given an angle $\theta$ whose terminal side does not lie along the $y$-axis, extend this terminal side, either outward or backward through the origin, until it intersects the auxiliary vertical line at a point $A$. Then the $y$-coordinate of $A$ is the value of $\tan \theta$. (The validity of this assertion is evident from Fig. 23; it depends on the correctness of the algebraic signs in all four quadrants, which we verify by inspection of the figure, and on the fact that

$$\frac{PQ}{OQ} = \frac{AB}{OB} = \frac{AB}{1} = AB.\Bigg)$$

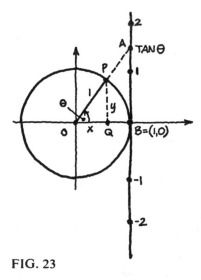

FIG. 23

All that remains is to sketch the graph by noticing that as $\theta$ in Fig. 23 increases from $-\frac{\pi}{2}$ to $\frac{\pi}{2}$ the point $A$ moves up the line from $-\infty$ through 0 to $+\infty$. This yields the middle part of Fig. 24, and the complete graph of $\tan \theta$ consists of infinitely many repetitions of this part to the right and to the left. It is clear that

$\tan (\theta + \pi) = \tan \theta,$

so the tangent function is periodic with period $\pi$.

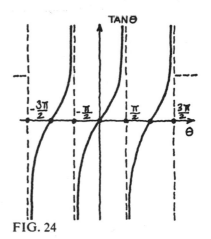

FIG. 24

## 7. THE MAJOR IDENTITIES

We now complete the list of fundamental identities started in Section 4. Let the Greek letters $\theta$ and $\phi$ (theta and phi) denote any two numbers. Our first three identities are called the *addition formulas:*

$$\sin (\theta + \phi) = \sin \theta \cos \phi + \cos \theta \sin \phi, \quad (12)$$
$$\cos (\theta + \phi) = \cos \theta \cos \phi - \sin \theta \sin \phi, \quad (13)$$
$$\tan (\theta + \phi) = \frac{\tan \theta + \tan \phi}{1 - \tan \theta \tan \phi}. \quad (14)$$

The first two of these are proved below, at the end of this section. The third follows from the first two by a relatively simple argument: first write

$$\tan (\theta + \phi) = \frac{\sin (\theta + \phi)}{\cos (\theta + \phi)}$$
$$= \frac{\sin \theta \cos \phi + \cos \theta \sin \phi}{\cos \theta \cos \phi - \sin \theta \sin \phi},$$

now divide both numerator and denominator of the fraction on the right by $\cos \theta \cos \phi$, obtaining

$$\tan (\theta + \phi) = \frac{\dfrac{\sin \theta}{\cos \theta} + \dfrac{\sin \phi}{\cos \phi}}{1 - \left(\dfrac{\sin \theta}{\cos \theta}\right)\left(\dfrac{\sin \phi}{\cos \phi}\right)},$$

which is (14).

The corresponding *subtraction formulas* are

$$\sin (\theta - \phi) = \sin \theta \cos \phi - \cos \theta \sin \phi, \quad (15)$$
$$\cos (\theta - \phi) = \cos \theta \cos \phi + \sin \theta \sin \phi, \quad (16)$$
$$\tan (\theta - \phi) = \frac{\tan \theta - \tan \phi}{1 + \tan \theta \tan \phi}. \quad (17)$$

These follow at once from the addition formulas by replacing $\phi$ by $-\phi$ and using (6), (7) and (8).

105

Thus, for example,

$$\sin (\theta - \phi) = \sin [\theta + (-\phi)]$$
$$= \sin \theta \cos (-\phi) + \cos \theta \sin (-\phi)$$
$$= \sin \theta \cos \phi - \cos \theta \sin \phi,$$

and similarly for the others.

The *double-angle formulas* are

$$\boxed{\begin{aligned} \sin 2\theta &= 2 \sin \theta \cos \theta, \\ \cos 2\theta &= \cos^2\theta - \sin^2\theta. \end{aligned}}$$

(18)
(19)

These are special cases of (12) and (13), obtained by replacing $\phi$ by $\theta$. There is, of course, an obvious double-angle formula for the tangent; but this is of minor significance and we therefore omit it.

The *half-angle formulas* are

$$\boxed{\begin{aligned} 2\cos^2\theta &= 1 + \cos 2\theta, \\ 2\sin^2\theta &= 1 - \cos 2\theta. \end{aligned}}$$

(20)
(21)

These identities are easy to establish by writing (9) and (19) together, as

$$\cos^2\theta + \sin^2\theta = 1,$$
$$\cos^2\theta - \sin^2\theta = \cos 2\theta.$$

By adding we get (20), and by subtracting we get (21).*

Our list of identities is now complete, and all of them have been proved except (12) and (13). We can establish these as follows for the special case in which both $\theta$ and $\phi$ are positive acute angles and their sum is also acute (Fig. 25). From any point $P$ on the terminal side of $\theta + \phi$ draw $PQ$ perpendicular to the $x$-axis and $PR$ perpen-

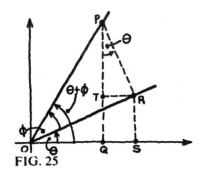

FIG. 25

---

*The name *half-angle formula* can be understood as follows. In (21), for instance, if $\theta$ is replaced throughout by $\frac{1}{2}\theta$, the result is $2 \sin^2 \frac{1}{2}\theta = 1 - \cos \theta$ or

$$\sin \frac{1}{2}\theta = \pm\sqrt{\frac{1 - \cos \theta}{2}}.$$

This version of (21) enables us to find the sine of half an angle if the cosine of the angle is known. Thus, for example,

$$\sin 15° = \sin \frac{\pi}{12} = \sin \frac{1}{2}\left(\frac{\pi}{6}\right) = \sqrt{\frac{1 - \frac{1}{2}\sqrt{3}}{2}} = \sqrt{\frac{2 - \sqrt{3}}{4}}$$
$$= \frac{1}{2}\sqrt{2 - \sqrt{3}}.$$

However, *the half-angle formulas find their main uses in calculus, and not in computations of this kind.*

dicular to the terminal side of $\theta$. From $R$, draw $RS$ perpendicular to the $x$-axis and $RT$ perpendicular to $PQ$. Notice that the angle $TPR$ equals $\theta$, since both are acute angles and their sides are respectively perpendicular. The proofs now go as follows:

$$\sin (\theta + \phi) = \frac{PQ}{OP} = \frac{PT + TQ}{OP} = \frac{PT + RS}{OP}$$

$$= \frac{PT}{OP} + \frac{RS}{OP}$$

$$= \frac{PT}{PR} \cdot \frac{PR}{OP} + \frac{RS}{OR} \cdot \frac{OR}{OP}$$

$$= \cos \theta \sin \phi + \sin \theta \cos \phi \cdot$$

and

$$\cos (\theta + \phi) = \frac{OQ}{OP} = \frac{OS - QS}{OP} = \frac{OS - TR}{OP}$$

$$= \frac{OS}{OP} - \frac{TR}{OP}$$

$$= \frac{OS}{OR} \cdot \frac{OR}{OP} - \frac{TR}{PR} \cdot \frac{PR}{OP}$$

$$= \cos \theta \cos \phi - \sin \theta \sin \phi.$$

These limited proofs may serve to convince most readers that (12) and (13) are indeed true. For those who wish to pursue this matter to the end, fully general proofs (of a somewhat different character) are provided in Appendix B.

# 8. THE INVERSE TRIGONOMETRIC FUNCTIONS

We know that $\sin \frac{\pi}{6} = \frac{1}{2}$. If, therefore, we are asked to find an angle (in radians) whose sine is $\frac{1}{2}$, we can answer immediately that $\frac{\pi}{6}$ is such an angle.

It is necessary in higher mathematics to have a symbol to denote an angle whose sine is a given number $x$. There are two such symbols in common use,

$$\sin^{-1}x \quad \text{and} \quad \arcsin x.$$

These symbols are fully equivalent and can be used interchangeably, though we shall confine ourselves to the first. The first is usually read as "the inverse sine of $x$" and the second as "the arc sine of $x$," and both mean "an angle whose sine is $x$." It is essential to understand that in the symbol $\sin^{-1}x$, the $-1$ is *not* an exponent, and therefore $\sin^{-1}x$ *never* means $\dfrac{1}{\sin x}$. The symbol $\sin^{-1}x$ is merely a short way of writing the expression "an angle whose sine is $x$."

The above statements can be summarized as follows: the formulas

$x = \sin y$   and   $y = \sin^{-1}x$

mean exactly the same thing, in the sense that

$$x = 2y \quad \text{and} \quad y = \frac{1}{2}x$$

mean exactly the same thing. In each case the equation is first written in a form solved for $x$, and then (the same equation!) in a form solved for $y$.

In order to sketch the graph of $y = \sin^{-1}x$, it suffices to sketch $x = \sin y$ and then flip the picture over, interchanging the positions of the axes (Fig. 26). It is evident that $y$ exists only when $x$ is confined to the interval $-1 \le x \le 1$. However, for any such $x$ there are infinitely many corresponding $y$'s, and this situation cannot be permitted if $y = \sin^{-1}x$ is to be considered a function. Recall that it is part of the meaning of the word "function" that to each $x$ there corresponds *only one* $y$. The standard way of dealing with this difficulty is to arbitrarily restrict the values of $y = \sin^{-1}x$ to the interval $-\dfrac{\pi}{2} \le y \le \dfrac{\pi}{2}$ (see the heavier portion of the curve in Fig. 26). The essence of the above explanation is this: we invent a new symbol, $y = \sin^{-1}x$, to signify the result of solving the equation $\sin y = x$ for $y$; and to make the solution single-valued, we specify that $y$ must lie in the interval $-\dfrac{\pi}{2} \le y \le \dfrac{\pi}{2}$.

The function $y = \tan^{-1} x$ (the equivalent notation here is $y = \arctan x$) is defined in almost the same way:

$y = \tan^{-1}x$   means   $\tan y = x$   and

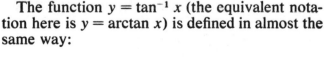

$$-\frac{\pi}{2} < y < \frac{\pi}{2}.$$

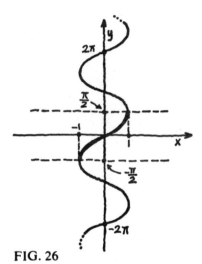

FIG. 26

The symbol $\tan^{-1}x$ is read "the inverse tangent of $x$," and it means "the angle (in the specified interval) whose tangent is $x$." The graph of $y = \tan^{-1}x$ is the heavy curve in Fig. 27.

It might seem that the other four trigonometric functions also have their inverses. And so they do; but these other inverse functions are totally unnecessary in the applications and will not be mentioned.

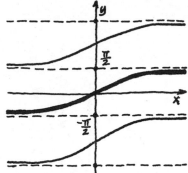

FIG. 27

## 9. THE LAW OF COSINES AND THE LAW OF SINES

The *law of cosines* is a very useful tool in a variety of situations in physics and geometry. It expresses the third side of a triangle (Fig. 28) in terms of two given sides $a$ and $b$ and the included angle $\theta$:

$$c^2 = a^2 + b^2 - 2ab\cos\theta.$$

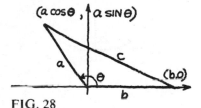

FIG. 28

The proof is very easy if we place the triangle in the $xy$-plane as shown in the figure and apply the distance formula to the vertices $(a\cos\theta, a\sin\theta)$ and $(b, 0)$. The square of the side $c$ is clearly

$$c^2 = (a\cos\theta - b)^2 + (a\sin\theta - 0)^2$$
$$= a^2(\cos^2\theta + \sin^2\theta) + b^2 - 2ab\cos\theta$$
$$= a^2 + b^2 - 2ab\cos\theta,$$

and the proof is complete.

The law of cosines has a companion which is much less important but which we include for the sake of completeness. If the three angles of a triangle and their opposite sides are labeled as shown in Fig. 29, then the *law of sines* states that

$$\frac{\sin A}{a} = \frac{\sin B}{b} = \frac{\sin C}{c}.$$

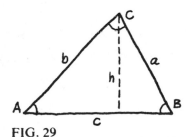

FIG. 29

To establish this, we insert the height $h$ of the triangle, as indicated, and use the two right triangles obtained in this way to write

$$\sin A = \frac{h}{b} \quad \text{and} \quad \sin B = \frac{h}{a},$$

which yield $h = b\sin A$ and $h = a\sin B$. Equating these two expressions for $h$, we get

$$b\sin A = a\sin B,$$

or equivalently,

$$\frac{\sin A}{a} = \frac{\sin B}{b}.$$

The remaining part of the law of sines is proved in the same way.

# APPENDIX A. THE BARE BONES OF THE SUBJECT

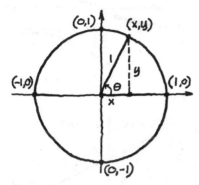

$\sin \theta = y$

$\cos \theta = x$

$\tan \theta = \dfrac{y}{x}$

$\cot \theta = \dfrac{x}{y}$

$\sec \theta = \dfrac{1}{x}$

$\csc \theta = \dfrac{1}{y}$

---

$\tan \theta = \dfrac{\sin \theta}{\cos \theta}$

$\cot \theta = \dfrac{\cos \theta}{\sin \theta}$

$\sec \theta = \dfrac{1}{\cos \theta}$

$\csc \theta = \dfrac{1}{\sin \theta}$

$\cot \theta = \dfrac{1}{\tan \theta}$

---

$\sin (-\theta) = -\sin \theta$

$\cos (-\theta) = \cos \theta$

$\tan (-\theta) = -\tan \theta$

---

$\sin^2\theta + \cos^2\theta = 1$

$\tan^2\theta + 1 = \sec^2\theta$

$1 + \cot^2\theta = \csc^2\theta$

---

$\sin (\theta + \phi) = \sin \theta \cos \phi + \cos \theta \sin \phi$

$\cos (\theta + \phi) = \cos \theta \cos \phi - \sin \theta \sin \phi$

$\tan (\theta + \phi) = \dfrac{\tan \theta + \tan \phi}{1 - \tan \theta \tan \phi}$

---

$\sin (\theta - \phi) = \sin \theta \cos \phi - \cos \theta \sin \phi$

$\cos (\theta - \phi) = \cos \theta \cos \phi + \sin \theta \sin \phi$

$\tan (\theta - \phi) = \dfrac{\tan \theta - \tan \phi}{1 + \tan \theta \tan \phi}$

$$\sin 2\theta = 2 \sin \theta \cos \theta$$
$$\cos 2\theta = \cos^2\theta - \sin^2\theta$$

$$2 \cos^2\theta = 1 + \cos 2\theta$$
$$2 \sin^2\theta = 1 - \cos 2\theta$$

Law of cosines:
$$c^2 = a^2 + b^2 - 2ab \cos \theta$$

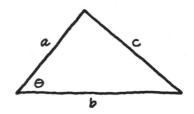

# APPENDIX B. COMPLETE PROOFS OF IDENTITIES (12) AND (13)

Let $\theta$ and $\phi$ be arbitrary angles with their initial sides lying along the positive $x$-axis, as shown in Fig. 30. Their terminal sides intersect the unit circle at the points $P = (\cos \theta, \sin \theta)$ and $Q = (\cos \phi, \sin \phi)$. By the distance formula the square of the distance $PQ$ is

$$\begin{aligned}
PQ^2 &= (\cos \theta - \cos \phi)^2 + (\sin \theta - \sin \phi)^2 \\
&= (\cos^2\theta + \sin^2\theta) + (\cos^2\phi + \sin^2\phi) \\
&\quad - 2(\cos \theta \cos \phi + \sin \theta \sin \phi) \\
&= 2 - 2(\cos \theta \cos \phi + \sin \theta \sin \phi).
\end{aligned}$$

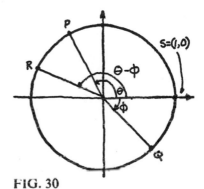

FIG. 30

Now rotate both points $P$ and $Q$ through the same angle $-\phi$—counterclockwise if $-\phi$ is positive and clockwise if $-\phi$ is negative—and observe that their new positions are the points $R = (\cos (\theta - \phi), \sin (\theta - \phi))$ and $S = (1, 0)$. The square of the distance $RS$ is

$$\begin{aligned}
RS^2 &= (\cos (\theta - \phi) - 1)^2 + (\sin (\theta - \phi) - 0)^2 \\
&= \cos^2(\theta - \phi) + \sin^2(\theta - \phi) + 1 \\
&\quad - 2 \cos (\theta - \phi) \\
&= 2 - 2 \cos (\theta - \phi).
\end{aligned}$$

However, the rotation does not change the distances between pairs of points, so $PQ = RS$, $PQ^2 = RS^2$, and by the above calculations we have identity (16) in Section 7:

$$\cos (\theta - \phi) = \cos \theta \cos \phi + \sin \theta \sin \phi. \qquad (16)$$

If we replace $\phi$ by $-\phi$ here, then we obtain identity (13):

$$\begin{aligned}
\cos (\theta + \phi) &= \cos [\theta - (-\phi)] \\
&= \cos \theta \cos (-\phi) + \sin \theta \sin (-\phi) \\
&= \cos \theta \cos \phi - \sin \theta \sin \phi. \qquad (13)
\end{aligned}$$

Now, replacing $\theta$ in (16) by $\frac{\pi}{2}$ yields

$$\cos\left(\frac{\pi}{2}-\phi\right) = \cos\frac{\pi}{2}\cos\phi + \sin\frac{\pi}{2}\sin\phi$$
$$= 0\cdot\cos\phi + 1\cdot\sin\phi$$
$$= \sin\phi;$$

and replacing $\phi$ by $\frac{\pi}{2}-\phi$ here yields

$$\sin\left(\frac{\pi}{2}-\phi\right) = \cos\left[\frac{\pi}{2}-\left(\frac{\pi}{2}-\phi\right)\right]$$
$$= \cos\phi.$$

These two formulas — and (16) — enable us to prove (12) at once:

$$\sin(\theta+\phi) = \cos\left[\frac{\pi}{2}-(\theta+\phi)\right]$$
$$= \cos\left[\left(\frac{\pi}{2}-\theta\right)-\phi\right]$$
$$= \cos\left(\frac{\pi}{2}-\theta\right)\cos\phi$$
$$\quad + \sin\left(\frac{\pi}{2}-\theta\right)\sin\phi$$
$$= \sin\theta\cos\phi + \cos\theta\sin\phi. \qquad (12)$$

We emphasize again that no restrictions whatever have been placed on the angles $\theta$ and $\phi$, so these proofs are completely general.

# APPENDIX C. A SHORT TABLE OF VALUES

| Angle | | Sine | Co-sine | Tan-gent | Angle | | Sine | Co-sine | Tan-gent |
|---|---|---|---|---|---|---|---|---|---|
| De-gree | Ra-dian | | | | De-gree | Ra-dian | | | |
| 0° | 0.000 | 0.000 | 1.000 | 0.000 | | | | | |
| 1° | 0.017 | 0.017 | 1.000 | 0.017 | 46° | 0.803 | 0.719 | 0.695 | 1.036 |
| 2° | 0.035 | 0.035 | 0.999 | 0.035 | 47° | 0.820 | 0.731 | 0.682 | 1.072 |
| 3° | 0.052 | 0.052 | 0.999 | 0.052 | 48° | 0.838 | 0.743 | 0.669 | 1.111 |
| 4° | 0.070 | 0.070 | 0.998 | 0.070 | 49° | 0.855 | 0.755 | 0.656 | 1.150 |
| 5° | 0.087 | 0.087 | 0.996 | 0.087 | 50° | 0.873 | 0.766 | 0.643 | 1.192 |
| 6° | 0.105 | 0.105 | 0.995 | 0.105 | 51° | 0.890 | 0.777 | 0.629 | 1.235 |
| 7° | 0.122 | 0.122 | 0.993 | 0.123 | 52° | 0.908 | 0.788 | 0.616 | 1.280 |
| 8° | 0.140 | 0.139 | 0.990 | 0.141 | 53° | 0.925 | 0.799 | 0.602 | 1.327 |
| 9° | 0.157 | 0.156 | 0.988 | 0.158 | 54° | 0.942 | 0.809 | 0.588 | 1.376 |
| 10° | 0.175 | 0.174 | 0.985 | 0.176 | 55° | 0.960 | 0.819 | 0.574 | 1.428 |
| 11° | 0.192 | 0.191 | 0.982 | 0.194 | 56° | 0.977 | 0.829 | 0.559 | 1.483 |
| 12° | 0.209 | 0.208 | 0.978 | 0.213 | 57° | 0.995 | 0.839 | 0.545 | 1.540 |
| 13° | 0.227 | 0.225 | 0.974 | 0.231 | 58° | 1.012 | 0.848 | 0.530 | 1.600 |
| 14° | 0.244 | 0.242 | 0.970 | 0.249 | 59° | 1.030 | 0.857 | 0.515 | 1.664 |
| 15° | 0.262 | 0.259 | 0.966 | 0.268 | 60° | 1.047 | 0.866 | 0.500 | 1.732 |
| 16° | 0.279 | 0.276 | 0.961 | 0.287 | 61° | 1.065 | 0.875 | 0.485 | 1.804 |
| 17° | 0.297 | 0.292 | 0.956 | 0.306 | 62° | 1.082 | 0.883 | 0.469 | 1.881 |
| 18° | 0.314 | 0.309 | 0.951 | 0.325 | 63° | 1.100 | 0.891 | 0.454 | 1.963 |
| 19° | 0.332 | 0.326 | 0.946 | 0.344 | 64° | 1.117 | 0.899 | 0.438 | 2.050 |
| 20° | 0.349 | 0.342 | 0.940 | 0.364 | 65° | 1.134 | 0.906 | 0.423 | 2.145 |
| 21° | 0.367 | 0.358 | 0.934 | 0.384 | 66° | 1.152 | 0.914 | 0.407 | 2.246 |
| 22° | 0.384 | 0.375 | 0.927 | 0.404 | 67° | 1.169 | 0.921 | 0.391 | 2.356 |
| 23° | 0.401 | 0.391 | 0.921 | 0.424 | 68° | 1.187 | 0.927 | 0.375 | 2.475 |
| 24° | 0.419 | 0.407 | 0.914 | 0.445 | 69° | 1.204 | 0.934 | 0.358 | 2.605 |
| 25° | 0.436 | 0.423 | 0.906 | 0.466 | 70° | 1.222 | 0.940 | 0.342 | 2.748 |
| 26° | 0.454 | 0.438 | 0.899 | 0.488 | 71° | 1.239 | 0.946 | 0.326 | 2.904 |
| 27° | 0.471 | 0.454 | 0.891 | 0.510 | 72° | 1.257 | 0.951 | 0.309 | 3.078 |
| 28° | 0.489 | 0.469 | 0.883 | 0.532 | 73° | 1.274 | 0.956 | 0.292 | 3.271 |
| 29° | 0.506 | 0.485 | 0.875 | 0.554 | 74° | 1.292 | 0.961 | 0.276 | 3.487 |
| 30° | 0.524 | 0.500 | 0.866 | 0.577 | 75° | 1.309 | 0.966 | 0.259 | 3.732 |
| 31° | 0.541 | 0.515 | 0.857 | 0.601 | 76° | 1.326 | 0.970 | 0.242 | 4.011 |
| 32° | 0.559 | 0.530 | 0.848 | 0.625 | 77° | 1.344 | 0.974 | 0.225 | 4.332 |
| 33° | 0.576 | 0.545 | 0.839 | 0.649 | 78° | 1.361 | 0.978 | 0.208 | 4.705 |
| 34° | 0.593 | 0.559 | 0.829 | 0.675 | 79° | 1.379 | 0.982 | 0.191 | 5.145 |
| 35° | 0.611 | 0.574 | 0.819 | 0.700 | 80° | 1.396 | 0.985 | 0.174 | 5.671 |
| 36° | 0.628 | 0.588 | 0.809 | 0.727 | 81° | 1.414 | 0.988 | 0.156 | 6.314 |
| 37° | 0.646 | 0.602 | 0.799 | 0.754 | 82° | 1.431 | 0.990 | 0.139 | 7.115 |
| 38° | 0.663 | 0.616 | 0.788 | 0.781 | 83° | 1.449 | 0.993 | 0.122 | 8.144 |
| 39° | 0.681 | 0.629 | 0.777 | 0.810 | 84° | 1.466 | 0.995 | 0.105 | 9.514 |
| 40° | 0.698 | 0.643 | 0.766 | 0.839 | 85° | 1.484 | 0.996 | 0.087 | 11.43 |
| 41° | 0.716 | 0.656 | 0.755 | 0.869 | 86° | 1.501 | 0.998 | 0.070 | 14.30 |
| 42° | 0.733 | 0.669 | 0.743 | 0.900 | 87° | 1.518 | 0.999 | 0.052 | 19.08 |
| 43° | 0.750 | 0.682 | 0.731 | 0.933 | 88° | 1.536 | 0.999 | 0.035 | 28.64 |
| 44° | 0.768 | 0.695 | 0.719 | 0.966 | 89° | 1.553 | 1.000 | 0.017 | 57.29 |
| 45° | 0.785 | 0.707 | 0.707 | 1.000 | 90° | 1.571 | 1.000 | 0.000 | |

# APPENDIX D. A FEW EXERCISES FOR THOSE WHO FEEL THE NEED OF THEM

The exercises with full solutions given in Appendix E are marked with an asterisk (*).

## SECTION 3

1. Express the given angles in radians: (a) 12°; (b) 24°; (c) 36°; (d) 15°; (e) 5°; (f) 20°; (g) 75°; (h) 80°; (i) 105°; (j) 270°; (k) 27°; (l) −720°; (m) 630°; (n) −240°; (o) 225°; (p) 285°; (q) 150°; (r) 450°.

2. Express the given angles in degrees: (a) $\frac{5\pi}{3}$; (b) $-\frac{2\pi}{3}$; (c) $\frac{5\pi}{4}$; (d) $\frac{7\pi}{4}$; (e) $\frac{\pi}{5}$; (f) $\frac{3\pi}{5}$; (g) $\frac{6\pi}{5}$; (h) $\frac{9\pi}{5}$; (i) $\frac{11\pi}{5}$; (j) $\frac{5\pi}{6}$; (k) $-\frac{13\pi}{6}$; (l) $\frac{7\pi}{6}$; (m) $\frac{4\pi}{9}$; (n) $\frac{5\pi}{12}$.

## SECTION 4

Establish the following identities.

1. $\dfrac{\sin\theta + \tan\theta}{\csc\theta + \cot\theta} = \sin\theta\tan\theta.$

2. $\dfrac{\sin\theta + \tan\theta}{1 + \sec\theta} = \sin\theta.$

*3. $\dfrac{1 - 2\cos^2\theta}{\sin\theta\cos\theta} = \tan\theta - \cot\theta.$

4. $\dfrac{\cot\theta + 1}{\cot\theta - 1} = \dfrac{1 + \tan\theta}{1 - \tan\theta}.$

5. $1 + \cot^2\theta = \dfrac{\sec^2\theta}{\sec^2\theta - 1}.$

6. $\dfrac{\cot\theta + 1}{\sin\theta + \cos\theta} = \csc\theta.$

7. $\dfrac{1 + \sec\theta}{\tan\theta} = \dfrac{\tan\theta}{\sec\theta - 1}.$

*8. $\dfrac{\sin\theta}{\sec\theta} = \dfrac{1}{\tan\theta + \cot\theta}.$

9. $(1 - \sin^2\theta)(1 + \tan^2\theta) = 1.$

10. $\dfrac{\tan\theta}{1 + \tan^2\theta} = \sin\theta\cos\theta.$

11. $\csc^2\theta - \cos^2\theta\,\csc^2\theta = 1.$

12. $\sin^4\theta - \cos^4\theta = 1 - 2\cos^2\theta.$

13. $\tan^2\theta \sin^2\theta - \cos^2\theta = \sec^2\theta - 2.$
14. $\tan \theta \csc \theta = \tan \theta \sin \theta + \cos \theta.$
15. $\cot^2\theta - \tan^2\theta = \csc^2\theta - \sec^2\theta.$
16. $\sec^2\theta + \csc^2\theta = \sec^2\theta \csc^2\theta.$
17. $\sec^2\theta \csc^2\theta = (\tan \theta + \cot \theta)^2.$
18. $\dfrac{1 + \cos \theta}{\sec \theta - \tan \theta} + \dfrac{\cos \theta - 1}{\sec \theta + \tan \theta} = 2 + 2\tan \theta.$

# SECTION 5

Find the numerical values of the following expressions.

1. $\dfrac{\sin \frac{\pi}{2} + \cos \frac{\pi}{2}}{\sin \pi + \cos \pi}.$

2. $\dfrac{1 + \tan^2\frac{\pi}{3}}{1 + \cot^2\frac{\pi}{3}}.$

3. $\dfrac{\sin \pi + \cos (-\pi)}{\sin \frac{\pi}{2} + \cos \left(-\frac{\pi}{2}\right)}.$

4. $\dfrac{\tan \frac{\pi}{3} + \tan \pi}{\cot \frac{\pi}{6} + \cot \frac{\pi}{2}}$

5. $\dfrac{\sin \pi \cos \pi \tan \pi}{\sin \frac{\pi}{3} \cos \frac{\pi}{3} \tan \frac{\pi}{3}}.$

6. $\dfrac{\sin \frac{3\pi}{2} - \cos \frac{5\pi}{2}}{\sin \frac{5\pi}{2} - \cos \frac{3\pi}{2}}.$

7. $\sin \frac{5\pi}{4} \sin \frac{3\pi}{4} \sin \frac{\pi}{4}.$

8. $\dfrac{\sin \frac{\pi}{3} \sin \frac{\pi}{2} - \cos \frac{\pi}{3} \cos \frac{\pi}{2}}{\sin \frac{\pi}{3} \cos \frac{\pi}{2} + \cos \frac{\pi}{3} \sin \frac{\pi}{2}}.$

# SECTION 7

1. Find a formula for *(a) $\sin 3\theta$ in terms of $\sin \theta$; (b) $\cos 3\theta$ in terms of $\cos \theta$; (c) $\cos 4\theta$ in terms of $\cos \theta$.

2. Show that
   (a) $\sin \theta \sin \phi$

   $$= \frac{1}{2}[\cos (\theta - \phi) - \cos (\theta + \phi)];$$

   (b) $\cos \theta \cos \phi$

   $$= \frac{1}{2}[\cos (\theta - \phi) + \cos (\theta + \phi)];$$

   (c) $\sin \theta \cos \phi$

   $$= \frac{1}{2}[\sin (\theta + \phi) + \sin (\theta - \phi)].$$

3. By writing $\alpha = \theta + \phi$ and $\beta = \theta - \phi$ in the preceding exercise ($\alpha$ and $\beta$ are the Greek letters alpha and beta), show that
   (a) $\sin \alpha + \sin \beta$

   $$= 2 \sin \frac{1}{2}(\alpha + \beta) \cos \frac{1}{2}(\alpha - \beta);$$

   (b) $\sin \alpha - \sin \beta$

   $$= 2 \cos \frac{1}{2}(\alpha + \beta) \sin \frac{1}{2}(\alpha - \beta);$$

   (c) $\cos \alpha + \cos \beta$

   $$= 2 \cos \frac{1}{2}(\alpha + \beta) \cos \frac{1}{2}(\alpha - \beta);$$

   (d) $\cos \alpha - \cos \beta$

   $$= -2 \sin \frac{1}{2}(\alpha + \beta) \sin \frac{1}{2}(\alpha - \beta).$$

4. Establish the following identities:
   (a) $\sin 2\theta = \dfrac{2 \tan \theta}{1 + \tan^2\theta}$;

   (b) $\cos 2\theta = \dfrac{1 - \tan^2\theta}{1 + \tan^2\theta}$;

   (c) $\tan \theta = \dfrac{\sin \theta + \sin 2\theta}{1 + \cos \theta + \cos 2\theta}$;

   (d) $\tan \theta = \dfrac{\sin 2\theta}{1 + \cos 2\theta}$;

   *(e) $\cot \theta = \dfrac{\sin 2\theta}{1 - \cos 2\theta}$;

   (f) $\tan^2\theta = \dfrac{1 - \cos 2\theta}{1 + \cos 2\theta}$;

   (g) $\tan \frac{1}{2}\theta = \dfrac{\sin \theta}{1 + \cos \theta}$;

   (h) $\tan \theta \tan \frac{1}{2}\theta = \sec \theta - 1$;

(i) $\dfrac{1 + \sin \theta + \cos \theta}{1 + \sin \theta - \cos \theta} = \cot \dfrac{1}{2}\theta;$

(j) $1 - 4 \sin^4\theta = \cos 2\theta(1 + 2 \sin^2\theta);$

(k) $\tan \dfrac{1}{2}\theta + \cot \dfrac{1}{2}\theta = 2 \csc \theta;$

(l) $\cos^4\theta = \dfrac{1}{8}(3 + 4\cos 2\theta + \cos 4\theta);$

(m) $\sin^4\theta = \dfrac{1}{8}(3 - 4\cos 2\theta + \cos 4\theta).$

# SECTION 8

Find the values of the following expressions.

1. $\sin^{-1}0.$
2. $\sin^{-1}1.$
3. $\sin^{-1}\left(-\dfrac{1}{2}\sqrt{3}\right).$
4. $\sin^{-1}\left(-\dfrac{1}{2}\right).$
5. $\tan^{-1}\dfrac{1}{3}\sqrt{3}.$
6. $\tan^{-1}1.$
7. $\tan^{-1}(-\sqrt{3}).$
8. $\tan^{-1}\left(-\dfrac{1}{3}\sqrt{3}\right).$
9. $\sin [\tan^{-1}(-1)].$
10. $\cos (\sin^{-1}1).$
11. $\sin \left(\sin^{-1}\dfrac{2}{3}\right).$
12. $\cos [2 \sin^{-1}(-1)].$
13. $\sec \left[2 \sin^{-1}\dfrac{1}{2}\sqrt{3}\right].$
14. $\cot \left[\tan^{-1}\dfrac{1}{3}\sqrt{3}\right].$
15. $\csc \left(\sin^{-1}\dfrac{1}{5}\right).$

# APPENDIX E. THE ANSWERS TO THE EXERCISES, WITH FULL SOLUTIONS FOR SOME

## SECTION 3

1. (a) $\dfrac{\pi}{15}$; (b) $\dfrac{2\pi}{15}$; (c) $\dfrac{\pi}{5}$; (d) $\dfrac{\pi}{12}$; (e) $\dfrac{\pi}{36}$; (f) $\dfrac{\pi}{9}$;

(g) $\frac{5\pi}{12}$; (h) $\frac{4\pi}{9}$; (i) $\frac{7\pi}{12}$; (j) $\frac{3\pi}{2}$; (k) $\frac{3\pi}{20}$; (l) $-4\pi$;

(m) $\frac{7\pi}{2}$; (n) $-\frac{4\pi}{3}$; (o) $\frac{5\pi}{4}$; (p) $\frac{19\pi}{12}$; (q) $\frac{5\pi}{6}$; (r) $\frac{5\pi}{2}$.

**2.** (a) 300°; (b) −120°; (c) 225°; (d) 315°; (e) 36°;
(f) 108°; (g) 216°; (h) 324°; (i) 396°; (j) 150°;
(k) −390°; (l) 210°; (m) 80°; (n) 75°.

## SECTION 4

**3.** To establish this, we work on the right side:

$$\tan\theta - \cot\theta = \frac{\sin\theta}{\cos\theta} - \frac{\cos\theta}{\sin\theta} = \frac{\sin^2\theta - \cos^2\theta}{\sin\theta\cos\theta}$$

$$= \frac{(1 - \cos^2\theta) - \cos^2\theta}{\sin\theta\cos\theta} = \frac{1 - 2\cos^2\theta}{\sin\theta\cos\theta}.$$

**8.** To establish this, we work on both sides:

$$\frac{\sin\theta}{\sec\theta} = \frac{\sin\theta}{\dfrac{1}{\cos\theta}} = \sin\theta\cos\theta;$$

$$\frac{1}{\tan\theta + \cot\theta} = \frac{1}{\dfrac{\sin\theta}{\cos\theta} + \dfrac{\cos\theta}{\sin\theta}} = \frac{1}{\dfrac{\sin^2\theta + \cos^2\theta}{\sin\theta\cos\theta}}$$

$$= \frac{\sin\theta\cos\theta}{\sin^2\theta + \cos^2\theta} = \sin\theta\cos\theta.$$

## SECTION 5

**1.** −1.  **2.** 3.  **3.** −1.  **4.** 1.  **5.** 0.  **6.** −1.

**7.** $-\frac{1}{4}\sqrt{2}$.  **8.** $\sqrt{3}$.

## SECTION 7

**1.** (a) $3\sin\theta - 4\sin^3\theta$, because

$\sin 3\theta = \sin(2\theta + \theta) = \sin 2\theta\cos\theta + \cos 2\theta\sin\theta$
$= (2\sin\theta\cos\theta)\cos\theta + (\cos^2\theta - \sin^2\theta)\sin\theta$
$= 2\sin\theta\cos^2\theta + \sin\theta\cos^2\theta - \sin^3\theta$
$= 3\sin\theta\cos^2\theta - \sin^3\theta$
$= 3\sin\theta(1 - \sin^2\theta) - \sin^3\theta$
$= 3\sin\theta - 4\sin^3\theta;$

(b) $4\cos^3\theta - 3\cos\theta$; (c) $8\cos^4\theta - 8\cos^2\theta + 1$.

**4.** (e) To establish this, we work on the right side:

$$\frac{\sin 2\theta}{1 - \cos 2\theta} = \frac{2\sin\theta\cos\theta}{2\sin^2\theta} = \frac{\cos\theta}{\sin\theta} = \cot\theta.$$

## SECTION 8

**1.** 0,  **2.** $\frac{\pi}{2}$.  **3.** $-\frac{\pi}{3}$.  **4.** $-\frac{\pi}{6}$.  **5.** $\frac{\pi}{6}$.  **6.** $\frac{\pi}{4}$.

# TRIGONOMETRY

**7.** $-\dfrac{\pi}{3}$. **8.** $-\dfrac{\pi}{6}$. **9.** $-\dfrac{1}{2}\sqrt{2}$. **10.** 0. **11.** $\dfrac{2}{3}$.

**12.** $-1$. **13.** $-2$. **14.** $\sqrt{3}$. **15.** 5.